Universitext

Universitext

Editors (North America): J.H. Ewing, F.W. Gehring, and P.R. Halmos

Aupetit: A Primer on Spectral Theory
Berger: Geometry I, II (two volumes)
Bliedtner/Hansen: Potential Theory
Booss/Bleecker: Topology and Analysis
Chandrasekharan: Classical Fourier Transforms
Charlap: Bieberbach Groups and Flat Manifolds
Chern: Complex Manifolds Without Potential Theory
Cohn: A Classical Invitation to Algebraic Numbers and Class Fields
Curtis: Abstract Linear Algebra
Curtis: Matrix Groups
van Dalen: Logic and Structure
Devlin: Fundamentals of Contemporary Set Theory
Edwards: A Formal Background to Mathematics I a/b
Edwards: A Formal Background to Mathematics II a/b
Emery: Stochastic Calculus
Fukhs/Rokhlin: Beginner's Course in Topology
Frauenthal: Mathematical Modeling in Epidemiology
Gardiner: A First Course in Group Theory
Gårding/Tambour: Algebra for Computer Science
Godbillon: Dynamical Systems on Surfaces
Goldblatt: Orthogonality and Spacetime Geometry
Humi/Miller: Second Course in Order Ordinary Differential Equations
Hurwitz/Kritikos: Lectures on Number Theory
Iverson: Cohomology of Sheaves
Kelly/Matthews: The Non-Euclidean Hyperbolic Plane
Kostrikin: Introduction to Algebra
Krasnoselskii/Pekrovskii: Systems with Hysteresis
Luecking/Rubel: Complex Analysis: A Functional Analysis Approach
Marcus: Number Fields
McCarthy: Introduction to Arithmetical Functions
Meyer: Essential Mathematics for Applied Fields
Mines/Richman/Ruitenburg: A Course in Constructive Algebra
Moise: Introductory Problem Course in Analysis and Topology
Montesinos: Classical Tesselations and Three Manifolds
Nikulin/Shafarevich: Geometries and Groups
Øksendal: Stochastic Differential Equations
Rees: Notes on Geometry
Reisel: Elementary Theory of Metric Spaces
Rey: Introduction to Robust and Quasi-Robust Statistical Methods
Rickart: Natural Function Algebras
Rotman: Galois Theory

(continued after index)

Bernard Aupetit

A Primer on
Spectral Theory

Springer-Verlag
New York Berlin Heidelberg London
Paris Tokyo Hong Kong Barcelona

B. Aupetit
Départment de Mathématiques
    et Statistique
Université Laval
Québec, Canada G1K 7P4

Mathematics Subject Classifications: Primary—46H, 46J, 46K, 47A; Secondary—31A, 32F

Printed on acid-free paper.

©1991 Springer-Verlag New York, Inc.

Camera-ready copy supplied by the author using TEX.

9 8 7 6 5 4 3 2 1

ISBN-13:978-0-387-97390-6   e-ISBN-13:978-1-4612-3048-9
DOI: 10.1007/978-1-4612-3048-9

Ce livre est dédié à la mémoire de mes parents.

Marcel Aupetit

Marie-Thérèse Le Charme

"Le cose passate fanno luce alle future, perché il mondo fu sempre di una medesima sorte, e tutto quello che è e sarà è stato in altro tempo; e le cose medesime ritornano ma sotto diversi nomi e colori; però ognuno non le riconosce, ma solo chi è savio e le osserva e considera diligentemente."

Francesco Guicciardini

PREFACE

This book grew out of lectures on spectral theory which the author gave at the Scuola Normale Superiore di Pisa in 1985 and at the Université Laval in 1987. Its aim is to provide a rather quick introduction to the new techniques of subharmonic functions and analytic multifunctions in spectral theory. Of course there are many paths which enter the large forest of spectral theory: we chose to follow those of subharmonicity and several complex variables mainly because they have been discovered only recently and are not yet much frequented. In our book *Propriétés spectrales des algèbres de Banach*, Berlin, 1979, we made a first incursion, a rather technical one, into these newly discovered areas. Since that time the bushes and the thorns have been cut, so the walk is more agreeable and we can go even further.

In order to understand the evolution of spectral theory from its very beginnings, it is advisable to have a look at the following books: Jean Dieudonné, *History of Functional Analysis*, Amsterdam, 1981; Antonie Frans Monna, *Functional Analysis in Historical Perspective*, Utrecht, 1973; and Frédéric Riesz & Béla Szökefalvi-Nagy, *Leçons d'analyse fonctionnelle*, Budapest, 1952. However the picture has changed since these three excellent books were written. Readers may convince themselves of this by comparing the classical textbooks of Frans Rellich, *Perturbation Theory*, New York, 1969, and Tosio Kato, *Perturbation Theory for Linear Operators*, Berlin, 1966, with the present work. They will discover that many of the results found in these books are direct consequences of results, very often more general, given in chapters III, V and VII.

The subharmonicity of the spectrum and the theory of analytic multifunctions also illustrate two striking facts in the history of mathematics. General spectral theory and the abstract theory of subharmonic functions were both invented by F. Riesz (see his *Oeuvres complètes*), but obviously, in the 1920s, he did not realize that the two theories are equivalent in the following sense. Given an analytic family $f(\lambda)$ of operators, the logarithm of the spectral radius of $f(\lambda)$ is subharmonic in

λ, and conversely, given a subharmonic function ϕ, then by Z. Słodkowski's result mentioned on page 167, there exists an analytic family of operators defined on ℓ^2 such that $\phi(\lambda)$ is equal to the logarithm of the spectral radius of $f(\lambda)$. Moreover, in trying to improve some results of F. Hartogs, K. Oka introduced in 1934 the notion of analytic multifunction, but obviously, he saw no link with spectral theory, and the notion remained dormant until the 1980s. So, even in mathematics, the words of the great Florentine statesman and historian Francesco Guicciardini are true.

The book is divided into seven chapters and an appendix. In the first chapter, we give a list of basic results in functional analysis without any proofs, except for those of Machado's theorem and Milman's theorem, which are rarely mentioned in the reference books. The results in this chapter will be used throughout the book.

The second chapter introduces the reader to some examples of bounded operators on Banach spaces and Hilbert spaces, and in particular gives F. Riesz's theory for compact operators and V.I. Lomonosov's theorem on invariant subspaces.

In the third chapter, we define and give examples of Banach algebras. The fundamental Holomorphic Functional Calculus Theorem is then proved. In §4, the subharmonic variation of the spectrum is given with many important applications, for instance the Spectral Maximum Principle (Theorem 3.4.13), Liouville's Spectral Theorem (Theorem 3.4.14), the Holomorphic Variation of Isolated Spectral Values (Theorem 3.4.20), the Subharmonicity of the n-th Diameter (Theorem 3.4.24), the Scarcity of Elements with Finite Spectrum (Theorem 3.4.25), and the Identity Principle for operators having spectra with zero as the only limit point.

The fourth chapter introduces the reader to I.M. Gelfand's theory of commutative Banach algebras and to the representation theory of non-commutative Banach algebras, with the standard results of N. Jacobson and A. Sinclair. At the end we give a new analytic proof of I. Kaplansky's theorem on locally algebraic operators, and we extend it to locally independent operators.

The fifth chapter, along with §4 of chapter III and with chapter VII, is the most important part of the book. It contains a great number of applications of spectral subharmonicity. The most striking applications are the generalization of B.E. Johnson's theorem on the uniqueness of the norm (Theorem 5.5.1), the Perturbation Theorem by Inessential Elements (Theorem 5.7.4), which immediately implies classical results of I.C. Gohberg, A.F. Ruston and B.A. Barnes, and the new subharmonic proof of B.A. Barnes's theorem on the existence of the socle (Theorem 5.7.8).

The sixth chapter is a classical presentation of the spectral theorem for normal operators on a Hilbert space, with a few applications at the end. In particular it includes the very nice extension of the Russo-Dye theorem obtained by L.T. Gardner. This chapter is independent of chapters V and VII, so it can be read just after chapter IV.

The seventh chapter is the most difficult one for two reasons. Firstly, it involves difficult mathematics, such as the theory of pseudoconvex sets. Secondly, our treatment is superficial. Nevertheless in spite of these drawbacks we have decided to include it in this book in order to induce the reader to learn more on this new subject which has (and certainly will have in the future) very important applications. In particular, the General Pełczyński Conjecture (Theorem 7.2.9) is solved, and applications to the distribution of spectral values in the plane (for instance Theorem 7.3.10) are given.

The appendix is a compendium of all the results needed in the theories of subharmonic functions, of capacity, and of functions of several complex variables. Obviously, no proofs are given because of the strict limit we fixed on the number of pages in our manuscript. They can be found in the standard textboooks mentioned.

Each chapter is followed by a list of problems. In some of them extra material is introduced. A few are very difficult to solve; in these cases we suggest that the reader take a look at the given references.

We have tried to keep the prerequisites to a strict minimum, and to develop the techniques of spectral subharmonicity and analytic multifunctions in the smallest number of pages. This is mainly because we believe that it is easier to learn something quickly and deeply in a small book than to learn in a big book. However the reader is assumed to be familiar with the matter which would generally be covered in one-semester courses on algebra, on complex analysis and on functional analysis. The material presented corresponds roughly to two semesters of lectures, supposing that the students are familiar with the basics of subharmonic functions, capacity and functions of several complex variables. Otherwise it would be better to start with an introductory course covering the matter in the appendix.

Firstly we would like to thank Churchill College, Cambridge, and the Scuola Normale Superiore di Pisa, for their hospitality during the academic year 1984–1985. There we began to write the manuscript of this book. We also thank the Natural Sciences and Engineering Research Council of Canada and the F.C.A.R. Fund of the Province of Quebec for their constant financial help which gave us the opportunity to travel and have interesting discussions with many mathematicians.

Obviously much thanks are due to the many attentive readers of the different versions of the manuscript, in particular to our friend Jaroslav Zemánek, to our former students Line Baribeau and Frédéric Gourdeau, and to our students Daniel Turcotte, Alain Fournier and Abdelaziz Maouche. Special thanks go to Louise Chamberland, Alain Charbonneau and Alain Fournier for the unpleasant work of typing this manuscript.

Quebec   Bernard Aupetit
February 1990

CONTENTS

Chapter I

SOME REMINDERS OF FUNCTIONAL ANALYSIS

This short chapter is intended as an aide-mémoire of some fundamental results in functional analysis that will be used in the rest of this book. Consequently we shall give no proofs as they can be easily found in all the standard textbooks (for instance [7], [8]), with two exceptions: the proof of Milman's theorem (Theorem 1.1.12) and the nice and simple proof of Machado's theorem (Theorem 1.1.16) which is a strong extension of the Stone-Weierstrass theorem.

THEOREM 1.1.1 (R. BAIRE).   *If X is a non-empty open subset of a complete metric space then the union of a countable collection of closed subsets of X with empty interiors also has an empty interior.*

This book will be concerned with vector spaces over the complex field.

THEOREM 1.1.2 (F. RIESZ).   *Every normed vector space containing a compact ball is finite-dimensional.*

THEOREM 1.1.3 (H.HAHN-S.BANACH).   *Suppose that Y is a subspace of a normed vector space X and f is a linear functional on Y such that*
$$|f(x)| \leq \|x\| , \quad \text{for } x \in Y.$$
Then f extends to a linear functional F on X such that
$$|F(x)| \leq \|x\| , \quad \text{for } x \in X.$$

COROLLARY 1.1.4.   *If X is a normed space and $a \in X$, there exists a linear functional F on X such that $F(a) = \|a\|$ and $|F(x)| \leq \|x\|$, for all $x \in X$.*

The following corollary is very useful in proving that an element of a Banach space can be approximated by elements of a subspace.

COROLLARY 1.1.5.   *Suppose that Y is a subspace of a normed space X and $a \in X$. Then a is not in the closure of Y if and only if there exists a bounded linear functional F on X such that $F(a) = 1$ and $F(x) = 0$, for all x in Y.*

Let X be a Banach space and X' denote its topological dual. Using the weak *-topology and Theorem 1.1.3 it is possible to get the following result, which will be used in Chapter IV, §1.

COROLLARY 1.1.6.   *Suppose that A and B are two disjoint non-empty convex and compact subsets of X', with respect to the weak*-topology. Then there exists $a \in X$ such that*

$$\sup\{\operatorname{Re} f(a) \colon f \in A\} < \inf\{\operatorname{Re} f(a) \colon f \in B\}.$$

Let K be a compact set and $C(K)$ denote the Banach space of continuous functions on K with complex values, equipped with the uniform norm $\|f\|_K = \sup_{x \in K} |f(x)|$. In that case F. Riesz has identified exactly the topological dual of $C(X)$ in the following manner. He first showed that a bounded linear functional on $C(K)$ can be written as a linear combination of normalized positive linear functionals F, normalized positive meaning that $F(f) \geq 0$ for $f \geq 0$ in $C(K)$ and $F(1) = 1$, or equivalently F satisfies $F(1) = 1 = \|F\|$ (see Theorem 6.2.16). He then gave a precise representation of normalized positive linear functionals.

THEOREM 1.1.7 (F. RIESZ REPRESENTATION THEOREM).   *Let K be a compact set and F be a normalized positive functional on $C(K)$. Then there exist a complete σ-algebra M on K containing all Borel sets and a unique probability measure μ on M which represents F in the sense that*

$$F(f) = \int_K f(x)\, d\mu(x)$$

for all $f \in C(K)$.

Conversely it is obvious that given such a probability measure μ on K the mapping $f \mapsto \int_K f(x)\, d\mu(x)$ defines a normalized positive linear functional on $C(K)$.

THEOREM 1.1.8 (S.BANACH-L.ALAOGLU).   *Let X be a Banach space and X' be its topological dual. Then the unit ball $U = \{F \colon F \in X', \|F\| \leq 1\}$ is weak*-compact in X'.*

THEOREM 1.1.9 (M.KREIN-D.MILMAN). *Suppose that X is a locally convex space. If K is a compact convex subset of X then it is the closed convex hull of its extreme points.*

In fact it is possible to be more precise. Given K a compact convex subset of X, we denote by $\text{ext}(K)$ the closure of the subset of extreme points of K. If μ is a probability measure on $\text{ext}(K)$, we shall say that $x \in K$ is *represented by* μ if

$$F(x) = \int_{\text{ext}(K)} F(y)\, d\mu(y)\,, \quad \text{for all } F \in X'.$$

By the analogue of Corollary 1.1.4 for locally convex linear spaces, x is the unique vector which can be represented by μ.

THEOREM 1.1.10. *Suppose that A is a compact subset of a locally convex space X. A point $x \in X$ is in the closed convex hull of A if and only if there exists a probability measure μ on A which represents x.*

PROOF. If μ represents x then $F(x) = \int_A F(y)\, d\mu(y)$, and as X is separated by its bounded linear functionals, x is in the closed convex hull of A. Conversely, by hypothesis, there exists a family in the convex hull of $\text{ext}(A)$ which converges to x, or equivalently there exist points $x_\alpha = \sum_{i=1}^{n_\alpha} \lambda_i^\alpha y_i^\alpha$ ($\lambda_i^\alpha > 0$, $\sum \lambda_i^\alpha = 1$, $y_i^\alpha \in \text{ext}(A)$, α in some directed set) which converge to x. Then x_α is represented by the probability measure $\mu_\alpha = \sum \lambda_i^\alpha \delta_{y_i^\alpha}$, where $\delta_{y_i^\alpha}$ denotes the Dirac measure concentrated at the point y_i^α. By Theorem 1.1.7 and Theorem 1.1.8 the set of all probability measures on $\text{ext}(A)$ may be identified with a weak *-compact convex subset of $C(\text{ext}(A))'$. So (μ_α) contains a subfamily converging to a probability measure μ on $\text{ext}(A)$. It is then easy to verify that μ represents x. $\square$

We now prove Milman's theorem, which will be used in Chapter IV, §1.

LEMMA 1.1.11. *Suppose that X is a locally convex space, that $x \in X$ and that K is a non-empty compact convex subset of X. Then x is an extreme point of K if and only if δ_x is the only probability measure on K which represents x.*

PROOF. Suppose that x is an extreme point of K and that the measure μ represents x. We have to prove that μ is supported by $\{x\}$. By regularity of μ it suffices to prove that $\mu(E) = 0$ for each compact set $E \subset K \setminus \{x\}$. Suppose that $\mu(E) > 0$ for such an E. From the compactness of E there exists $y \in E$ such that $\mu(U \cap K) > 0$ for every neighbourhood U of y. Let U be a closed convex neighbourhood of y such that $U \cap K \subset K \setminus \{x\}$. The set $U \cap K$ is compact and convex and $0 < r = \mu(U \cap K) < 1$, for otherwise $\mu(U \cap K) = 1$, which would imply that $x \in U \cap K$ as x is represented by μ. Thus we define two new Borel measures on K by

$$\begin{cases} \mu_1(B) = \dfrac{1}{r}\,\mu(B \cap U \cap K) \\[2mm] \mu_2(B) = \dfrac{1}{1-r}\,\mu(B \setminus (U \cap K)) \end{cases}$$

for each Borel subset B of K. Let x_i be the unique vector which is represented by μ_i. Then $\mu_1(U \cap K) = 1$ implies $x_1 \in U \cap K$, so $x_1 \neq x$. Furthermore $\mu = r\mu_1 + (1-r)\mu_2$ implies that $x = rx_1 + (1-r)x_2$, a contradiction. The converse is easy to prove (see Exercise I.6). $\Box$

THEOREM 1.1.12 (D. MILMAN). *Suppose that K is a compact convex subset of a locally convex space, that $L \subset K$ and that K is the closed convex hull of L. Then the extreme points of K are contained in the closure of L.*

PROOF. Let $x \in \text{ext}(K)$. By Theorem 1.1.10 applied to $A = \overline{L}$, there exists a measure μ on $\overline{L}$ which represents x. By Lemma 1.1.11 we have $\mu = \delta_x$, so $x \in \overline{L}$. $\Box$

THEOREM 1.1.13 (S.BANACH-H.STEINHAUS). *Let X, Y be two Banach spaces and let $(T_\alpha)_{\alpha \in \Lambda}$ be a family of bounded linear mappings from X into Y. Suppose that for each $x \in X$ we have*

$$\sup_{\alpha \in \Lambda} \|T_\alpha x\| < +\infty.$$

Then $\sup_{\alpha \in \Lambda} \|T_\alpha\| < +\infty$.

THEOREM 1.1.14 (OPEN MAPPING THEOREM). *Let X, Y be two Banach spaces and let T be a bounded linear mapping from X onto Y. Then T is open. Moreover if T is injective then T^{-1} is a bounded linear mapping and there exist $\alpha, \beta > 0$ such that $\alpha\|x\| \leq \|Tx\| \leq \beta\|x\|$, for all $x \in X$.*

The next result is very important. It will be used often to prove continuity of a linear mapping.

THEOREM 1.1.15 (CLOSED GRAPH THEOREM). *Let X, Y be two Banach spaces and let T be a linear mapping from X into Y. Let $G = \{(x, Tx) : x \in X\} \subset X \times Y$, denote the graph of T. Then T is bounded if and only if G is closed in $X \times Y$.*

Hence in order to prove that T is bounded it suffices to prove the following property: if (x_n) converges to 0 and (Tx_n) converges to $a \in Y$ then $a = 0$.

If K is a real segment $[a, b]$, K. Weierstrass proved that every f in $C([a, b])$ may be uniformly approximated by polynomials on $[a, b]$. This result was generalized by the Stone-Weierstrass theorem which is an essential tool in functional analysis (for instance it will be used in the proof of Theorem 6.2.6). There are many proofs of this theorem, including the original one by M.H. Stone, and also I. Glicksberg's argument. The latter is based on an ingenious idea due to L. de Branges, which is short but requires several tools from functional analysis such as the Hahn-Banach theorem, Krein-Milman theorem, Banach-Alaoglu theorem and the Riesz representation theorem (this last proof is given in [10], pp. 5–7).

We now intend to give a very simple proof due to S. Machado. However, its limitation is that it uses the axiom of choice or equivalently Zorn's lemma. (Obviously, if K is separable, Zorn's lemma is not necessary.)

Let A be a closed subalgebra of $C(K)$ containing the constant 1. A non-empty subset E of K is said to be *A-antisymmetric* if whenever $h \in A$ and h is real on E then h is constant on E.

Suppose moreover that A is self-adjoint, that is $h \in A$ implies $\overline{h} \in A$, and separates the points of K, that is if $x, y \in K$ with $x \neq y$ there exists $h \in A$ such that $h(x) \neq h(y)$. Then the only A-antisymmetric sets are singletons. The reason is simple: suppose that two different points x, y are in the antisymmetric set E. Then there exists $h \in A$ such that $h(x) \neq h(y)$, so $\operatorname{Re} h = \frac{h+\overline{h}}{2}$ or $\operatorname{Im} h = \frac{h-\overline{h}}{2i}$ separates x and y. As these two functions are real-valued and in A, we get a contradiction.

If $f \in C(K)$ and F is a non-empty closed subset of K we introduce the following notation:

$$\|f\|_F = \sup_{x \in F} |f(x)|, \quad d_f(F) = \inf_{g \in A} \|f - g\|_F.$$

Obviously we have $d_f(F_1) \leq d_f(F_2)$ if $F_1 \subset F_2$.

THEOREM 1.1.16 (S. MACHADO). *Let $f \in C(K)$ and let A be a closed subalgebra of $C(K)$ containing the constant 1. Then there exists a closed A-antisymmetric subset E of K such that $d_f(E) = d_f(K)$.*

PROOF. Let $\mathcal{F}$ be the family of all non-empty closed subsets F of K such that $d_f(F) = d_f(K)$. Obviously $K \in \mathcal{F}$. If $\mathcal{C}$ is a subfamily of $\mathcal{F}$ totally ordered by inclusion then $G = \bigcap_{F \in \mathcal{C}} F$ is also in $\mathcal{F}$ for the following reasons. Given $g \in A$, we have $d_f(F) = d_f(K)$ for all $F \in \mathcal{C}$, so the sets $\{x : x \in F, |f(x) - g(x)| \geq d_f(K)\}$ are compact and non-empty. Thus their intersection $\{x : x \in G, |f(x) - g(x)| \geq d_f(K)\}$ is also compact and non-empty, which implies that $d_f(G) = d_f(K)$. By Zorn's lemma there exists a minimal element E in $\mathcal{F}$. We now intend to prove that E is A-antisymmetric.

Suppose this is false. There exists $h \in A$ which is both real and non-constant on E. Replacing h by a linear combination of h and 1 we may suppose that $\min_{x \in E} h(x) = 0$ and $\max_{x \in E} h(x) = 1$. Let us define

$$E_1 = \left\{x : x \in E, 0 \leq h(x) \leq \frac{2}{3}\right\}, \quad E_2 = \left\{x : x \in E, \frac{1}{3} \leq h(x) \leq 1\right\}.$$

Since E_1 and E_2 are non-empty proper closed subsets of E, the minimality of E implies that there exist $g_1, g_2 \in A$ such that

$$\|f - g_1\|_{E_1} < d_f(K) \quad \text{and} \quad \|f - g_2\|_{E_2} < d_f(K). \tag{1}$$

For $n \geq 1$, let $h_n = (1 - h^n)^{2^n}$ and $k_n = h_n \cdot g_1 + (1 - h_n) \cdot g_2$. We have $h_n, k_n \in A$ amd $0 \leq h_n \leq 1$ on E. From (1) it follows that

$$\begin{aligned}
\|f - k_n\|_{E_1 \cap E_2} &\leq h_n \|f - g_1\|_{E_1 \cap E_2} + (1 - h_n)\|f - g_2\|_{E_1 \cap E_2} \\
&\leq h_n \|f - g_1\|_{E_1} + (1 - h_n)\|f - g_2\|_{E_2} < d_f(K).
\end{aligned} \tag{2}$$

On $E_1 \setminus E_2$, where $0 \leq h(x) < \frac{1}{3}$, we have:

$$h_n \geq 1 - 2^n h^n \geq 1 - \left(\frac{2}{3}\right)^n. \tag{3}$$

On $E_2 \setminus E_1$, where $\frac{2}{3} \leq h(x) \leq 1$, we have:

$$h_n \leq (1 + h^n)^{-2^n} \leq 1/2^n h^n \leq \left(\frac{3}{4}\right)^n. \tag{4}$$

So k_n converges uniformly to g_1 on $E_1 \setminus E_2$ and uniformly to g_2 on $E_2 \setminus E_1$. Using (1) and (2) we conclude that $\|f - k_n\|_E < d_f(K)$ for n large enough, so $d_f(E) < d_f(K)$, which is a contradiction. $\square$

COROLLARY 1.1.17 (M.H.STONE-K.WEIERSTRASS). *Let K be a compact set and let A be a closed self-adjoint subalgebra of $C(K)$ containing the constant functions and separating the points of K. Then $A = C(K)$.*

PROOF. Let $f \in C(K)$. By Theorem 1.1.16 there exists a closed A-antisymmetric set E such that $d_f(E) = d_f(K)$. But we saw previously that E is a singleton $\{a\}$, so considering $g(x) = f(a)\mathbf{1}$ we conclude that $d_f(E) = 0$. Hence $d_f(K) = 0$ which means that $f \in A$, since A is closed. $\square$

Let K be a compact metric space. We say that $\mathcal{F} \subset C(K)$ is an *equicontinuous family* if there exist $C > 0$ and $\alpha > 0$ such that $|f(x) - f(y)| \leq Cd(x,y)^{\alpha}$, for all $x, y \in K$ and $f \in \mathcal{F}$, where d denotes the distance on K.

Relatively compact subsets of $C(K)$ can be easily characterized by the following result.

THEOREM 1.1.18 (C.ARZELÀ-G.ASCOLI). *Let K be a compact metric space. Then $\mathcal{F}$ is relatively compact in $C(K)$ if and only if it satisfies the following two conditions:*

(i) $\mathcal{F}$ is equicontinuous,

(ii) $\sup_{f \in \mathcal{F}} |f(x)| < +\infty$, for each $x \in K$.

EXERCISE 1.   Let f be a continuous function on $[0,1]$ with real values.

(i) First suppose that f is differentiable at a point $a \in [0,1]$. Prove that there exists an integer $n \geq 1$ such that $|f(x) - f(a)| \leq n|x - a|$, for all $x \in [0,1]$.

(ii) Let $E_n = \{f : f \in C([0,1])$, for which there exists some a in $[0,1]$, depending on f, such that $|f(x) - f(a)| \leq n|x - a|$, for all $x \in [0,1]\}$. Prove that E_n is closed and has no interior point in $C([0,1])$ for the topology defined by the uniform norm.

(iii) Derive from that the existence of a continuous function on $[0,1]$ which is not differentiable at every point of $[0,1]$.

EXERCISE 2.   Let X be a non-empty subset of a complete metric space. Let (f_n) be a sequence of complex functions defined on X such that the $|f_n|$ are lower semicontinuous on X and the f_n converge pointwise on X. Prove that there exist a ball $B \subset X$ and $C > 0$ such that $\sup_{x \in B} |f_n(x)| \leq C$ (W.F. Osgood's lemma).

EXERCISE 3.   Let X be a vector space and let A, B be two disjoint convex subsets of X. Using Zorn's lemma prove that there exist two disjoint convex sets A_0, B_0 such that $A \subset A_0$, $B \subset B_0$ and $X = A_0 \cup B_0$ (S. Kakutani's theorem).

EXERCISE 4.   Let X be a normed vector space and let A, B be two disjoint convex subsets of X. Suppose that A has interior points. Prove that there exists a bounded linear functional f such that $\sup_{x \in A} f(x) \leq \inf_{x \in B} f(x)$.

EXERCISE 5.   A $n \times n$ matrix (a_{ij}) is *double stochastic* if $a_{ij} \geq 0$, $\sum_{j=1}^{n} a_{ij} = 1$, for all i and $\sum_{i=1}^{n} a_{ij} = 1$, for all j. It is called a *permutation matrix* if, moreover, the entries a_{ij} take only the values 0 and 1. Using the Krein-Milman theorem prove that a double stochastic matrix is a convex combination of permutation matrices.

EXERCISE 6.   As mentioned in Lemma 1.1.11, if δ_x is the only probability measure representing x, prove that x is an extreme point.

EXERCISE 7.   Let $K_1, \ldots, K_n$ be n compact sets and $K = K_1 \times \ldots \times K_n$. Prove that every $f \in C(K)$ can be uniformly approximated on K by finite sums of expressions

$$f_1(x_1) \times \ldots \times f_n(x_n)$$

where $f_1 \in C(K_1), \ldots, f_n \in C(K_n)$.

EXERCISE 8.   Let K be a locally compact topological space. By $C_0(K)$ we denote the algebra of continuous functions on K vanishing at infinity, that is $f \in C_0(K)$

if and only if $\{x : x \in K, |f(x)| \geq \epsilon\}$ is compact for all $\epsilon > 0$. Extend the Stone-Weierstrass theorem to $C_0(K)$.

*EXERCISE 9.   Let $f \in C_0([0, +\infty])$. Prove that it can be uniformly approximated on $[0, +\infty]$ by functions of the form $e^{-\alpha x}p(x)$, where $\alpha > 0$ and p is a polynomial.

*EXERCISE 10.   Let $f \in C_0([-\infty, +\infty])$. Prove that it can be uniformly approximated on $[-\infty, +\infty]$ by functions of the form $e^{-\alpha x^2}p(x)$, where $\alpha > 0$ and p is a polynomial.

*EXERCISE 11.   Let $\mathcal{F}$ be a family of holomorphic functions defined on a domain U of the complex plane. We say that $\mathcal{F}$ is a *normal family* if every sequence of elements of $\mathcal{F}$ contains a subsequence which converges uniformly on compact subsets of U. Using Cauchy's inequalities and the Arzelà-Ascoli theorem, prove that $\mathcal{F}$ is normal if and only if it is uniformly bounded on each compact subset of U (P. Montel's theorem).

Chapter II

SOME CLASSES OF OPERATORS

§1. Finite-Dimensional Operators

Let X be a finite-dimensional vector space. We know that all norms on X are equivalent and this implies in particular that all linear mappings T from X into X are continuous. Indeed if $\| \cdot \|$ is a norm on X and if $e_1, \ldots, e_n$ is a basis of X, then we have

$$\|Tx\| \leq (|\lambda_1| + \cdots + |\lambda_n|) \max_{i=1,\ldots,n} \|Te_i\| , \quad \text{for } x = \lambda_1 e_1 + \cdots + \lambda_n e_n .$$

As $\|x\|_1 = |\lambda_1| + \cdots + |\lambda_n|$ is a norm on X which is equivalent to $\| \cdot \|$, we have the result.

Of course, using matrices, it is possible to build all the linear mappings from X into X. So the algebra $\mathcal{L}(X)$ of all linear mappings from X into X is isomorphic to $M_n(\mathbb{C})$, the isomorphism depending on the choice of the basis.

The theory of finite-dimensional operators or *matrix theory* is very well-known. There are many classical textbooks on this subject (for instance P. Lancaster and M. Tismenetsky, *The Theory of Matrices*, Orlando, 1985), so we only intend to give a brief survey of the properties of matrices.

Every matrix can be written as the sum of a diagonal one and a nilpotent one (triangularization of matrices). This decomposition can be even much more precise (Jordan decomposition in diagonal blocks). If a matrix T is self-adjoint ($T = {}^t\overline{T}$) or normal ($T^t\overline{T} = {}^t\overline{T}T$) then it can be diagonalized. In other words, it means that there exist $k \leq n$ orthogonal projections $P_1, \ldots, P_n$, satisfying $P_i^2 = P_i$, $P_i = {}^t\overline{P}_i$, $P_iP_j = 0$ for $i \neq j$ and $I = P_1 + \cdots + P_k$, such that

$$T = \lambda_1 P_1 + \cdots + \lambda_k P_k.$$

These projections are the orthogonal projections corresponding to the different eigenspaces of T. This result will be generalized for self-adjoint or normal compact operators on a Hilbert space in §3, and even for arbitrary self-adjoint and normal operators on a Hilbert space in Chapter VI.

From the Cayley-Hamilton theorem, the algebra $M_n(\mathbb{C})$ is algebraic in the sense that we have $p(T) = 0$ for all $T \in M_n(\mathbb{C})$ where $p(\lambda) = \det(T - \lambda I)$. In Chapter III, §3 we shall give a simple and nice analytic proof of this fundamental theorem. The algebra $M_n(\mathbb{C})$ has another interesting property:

THEOREM 2.1.1.   *The algebra $M_n(\mathbb{C})$ is simple, that is, its only two-sided ideals are $\{0\}$ and $M_n(\mathbb{C})$.*

PROOF.   Let e_{ij} be the matrix having every coefficient equal to 0 except at the intersection of the ith row and the jth column where it is 1. It is easy to verify that $e_{ij}e_{km} = 0$ if $j \neq k$ and $e_{ij}e_{jm} = e_{im}$. Let I be a non-zero two-sided ideal of $M_n(\mathbb{C})$ and $a \neq 0$, $a \in I$. Then $a = \sum_{i,j=1}^{n} \alpha_{ij}e_{ij}$. Suppose that $\alpha_{qp} \neq 0$. Then $e_{pq}ae_{pq} = e_{pq}\sum_{i=1}^{n}\alpha_{ip}e_{iq} = \alpha_{qp}e_{pq}$. Hence $e_{pq} \in I$. Then $e_{ij} = e_{ip}e_{pq}e_{qj} \in I$. Consequently $I = M_n(\mathbb{C})$. ∎

All finite-dimentional algebras over the complex field have a very simple structure. For instance if A is semi-simple (see Chapter III, §1 for the definitions of Jacobson radical and semi-simple algebras) we have the very famous:

THEOREM 2.1.2 (J.H.M.WEDDERBURN-E.ARTIN).   *Let A be a semi-simple finite-dimensional algebra over $\mathbb{C}$. Then there exist integers $n_1, \ldots, n_k \geq 1$ such that $A \simeq M_{n_1}(\mathbb{C}) \oplus \cdots \oplus M_{n_k}(\mathbb{C})$.*

PROOF.   See for instance I. Herstein, *Noncommutative Rings*, 1968, or Lemma 5.4.1. ∎

This result implies, in particular, that a commutative, semi-simple and finite-dimensional complex algebra is isomorphic to $\mathbb{C}^n$, where n denotes its dimension (See Exercise II.1).

If $T \in M_n(\mathbb{C})$, what can be said about the multifunction $T \mapsto \text{Sp}\,T = \{\lambda : \det(T - \lambda I) = 0\}$? Using the Implicit Function Theorem it can be shown that it depends continuously on T (this will in fact be a corollary of a much more general result, see Corollary 3.4.5). But is it analytic in some sense? For instance if F is an analytic function from $\mathbb{C}$ into $M_n(\mathbb{C})$, that is a family of matrices with entries depending analytically on a parameter λ, are the elements of $\operatorname{Sp} F(\lambda)$ analytic functions of λ? This is false in general. Consider for instance $F(\lambda) = \begin{pmatrix} 0 & \lambda \\ 1 & 0 \end{pmatrix} \in M_2(\mathbb{C})$ for which $\operatorname{Sp} F(\lambda) = \{\sqrt{\lambda}, -\sqrt{\lambda}\}$.

Nevertheless, once more using the Implicit Function Theorem, it is possible to prove that for λ outside of an exceptional set, called the set of *branching points*, the points of $\operatorname{Sp} F(\lambda)$ vary holomorphically. In the previous example there is only one branching point at 0. In fact, this result is a corollary of a much more general result (see Theorem 3.4.25).

Such functions $\lambda \mapsto \operatorname{Sp} F(\lambda)$ defined on a domain $D \subset \mathbb{C}$ can be identified with multifunctions, which are sometimes called *algebroid multifunctions*. These are functions $\lambda \mapsto K(\lambda)$ with $\lambda \in D$ defined by

$$K(\lambda) = \{z : z^n + a_1(\lambda)z^{n-1} + \cdots + a_n(\lambda) = 0\},$$

where $a_1, \ldots, a_n$ are holomorphic on D.

Such functions were studied for the first time by C. Puiseux around 1850. They are of great importance for algebraic geometry and the theory of Riemann surfaces.

This notion of an algebroid multifunction will be generalized in Chapter VII.

§2. Bounded Linear Operators on a Banach Space

It is possible to construct a great number of explicit linear operators on classical Banach spaces like ℓ^2, L^2, ℓ^p, L^p, c_0, L^∞, $C(K)$ etc. Hilbert spaces are the most interesting examples because of their nice geometrical properties. But what happens in the case of a general Banach space? Certainly it contains all the multiples of the identity operator I, but are there any other linear operators?

It is easy to build operators having finite or cofinite rank. Let X be a Banach space of infinite dimension and let E be a finite-dimensional subspace of X having basis $e_1, \ldots, e_n$ (it is automatically closed in X). Let $f_1, \ldots, f_n$ be n given bounded linear functionals on X. Then the operator T defined by

$$Tx = f_1(x)e_1 + \cdots + f_n(x)e_n$$

is bounded and its range is included in E. Moreover if we suppose that $f_i(e_j)$ is one if $i = j$ and zero if $i \neq j$ (this is possible by Theorem 1.1.3) then $f_i(x - Tx) = 0$ for $i = 1, \ldots, n$ and the linear operator $I - T$ has its range equal to $\operatorname{Ker} f_1 \cap \cdots \cap \operatorname{Ker} f_n$, and so it has finite codimension. In fact it is possible to go further.

LEMMA 2.2.1. *Let $a_1,\ldots,a_n$ be n linearly independent vectors in a normed vector space X. Then there exists $\epsilon > 0$ such that if $b_1,\ldots,b_n \in X$ satisfy $\max_{i=1,\ldots,n} \|b_i\| < \epsilon$ then $a_1+b_1,\ldots,a_n+b_n$ are linearly independent.*

PROOF. For $n = 1$ it is obvious. Supposing the result to be true for $n-1$, we shall prove it for n. Suppose that there exist sequences $(b_1^k),\ldots,(b_n^k)$ converging to 0 in X and sequences of complex numbers $(\alpha_1^k),\ldots,(\alpha_{n-1}^k)$ such that

$$a_n + b_n^k = \alpha_1^k(a_1 + b_1^k) + \cdots + \alpha_{n-1}^k(a_{n-1} + b_{n-1}^k).$$

If all the sequences (α_i^k) are bounded, applying the Bolzano-Weierstrass theorem, they contain converging subsequences and so $a_1,\ldots,a_n$ are linearly dependent, which is a contradiction. Without loss of generality suppose for instance that $\lim_k |\alpha_1^k| = +\infty$. Then we have:

$$(a_1 + c_1^k) + \beta_2^k(a_2 + b_2^k) + \cdots + \beta_{n-1}^k(a_{n-1} + b_{n-1}^k) = 0$$

with $\beta_i^k = \alpha_i^k/\alpha_1^k$ for $i = 2,\ldots,n-1$ and $c_1^k = b_1^k - \frac{a_n+b_n^k}{\alpha_1^k}$. But $\lim_k c_1^k = 0$. So by the induction hypothesis we get a contradiction. The lemma is proved. $\square$

THEOREM 2.2.2. *Let X be an infinite-dimensional Banach space. Then there exists a bounded linear operator on X having an infinite-dimensional range which is a limit of finite-rank operators.*

PROOF. Let us denote by $\mathfrak{F}$ the linear subspace of finite-rank operators, by $\overline{\mathfrak{F}}$ its closure for the norm topology and by $\mathfrak{F}_n$ the set of operators with rank $\leq n$. Suppose the theorem false. Then $\overline{\mathfrak{F}} = \cup_{n=0}^\infty \mathfrak{F}_n$. But $\mathfrak{F}_n$ is closed for the following reasons. Suppose that $T \in \overline{\mathfrak{F}}_n$ and $T \notin \mathfrak{F}_n$. Then there exists a sequence (T_k) converging to T with $T_k \in \mathfrak{F}_n$ for $k = 1,2,\ldots$ and there exist $n+1$ unit vectors $x_1,\ldots,x_{n+1} \in X$ such that $Tx_1,\ldots,Tx_{n+1}$ are linearly independent. If we choose k large enough such that $\|T - T_k\| < \epsilon$ for the ϵ obtained in Lemma 2.2.1 applied to the $n+1$ vectors $a_1 = Tx_1,\ldots,a_{n+1} = Tx_{n+1}$, we get a contradiction because the $T_kx_1,\ldots,T_kx_{n+1}$ are linearly dependent. So $\mathfrak{F}_n$ is closed. The linear subspace $\overline{\mathfrak{F}}$ is complete so, by Theorem 1.1.1, one of the $\mathfrak{F}_n$, for instance $\mathfrak{F}_N$, contains a ball of $\overline{\mathfrak{F}}$ with centre S and radius $r > 0$. Let $T \in \mathfrak{F}$ be arbitrary. Then S and $S + \lambda T$ are in $\mathfrak{F}_N$ for λ small. Consequently T has rank $\leq 2N$. Using the construction we gave at the beginning of §2 we can build operators of arbitrary rank, so we get a contradiction. $\square$

It is easy to see that $\mathfrak{F}$ is a two-sided ideal of $\mathfrak{L}(X)$, the algebra of bounded linear mappings from X into X, and consequently $\overline{\mathfrak{F}}$ is a closed two-sided ideal of $\mathfrak{L}(X)$. If $S \in \mathfrak{F}$ and if B denotes the closed unit ball of X then $\overline{S(B)}$ is closed and bounded in the range of S, so it is compact in this range and also in all X. Let us show that this property is also true for $T \in \overline{\mathfrak{F}}$. First we need an elementary lemma in topology.

LEMMA 2.2.3. *Let F be a closed subset of a complete metric space. Then the following conditions are equivalent:*

(i) *F is compact,*

(ii) *F has the ϵ-covering property for all $\epsilon > 0$, that is for every $\epsilon > 0$ there exist finitely many balls of radius ϵ covering F.*

PROOF. It is obvious that (i) implies (ii). So we suppose that (ii) is true. Let $\mathcal{U}$ be an open covering of F. To reach a contradiction, suppose that no finite subcollection of $\mathcal{U}$ covers F. By hypothesis F is the union of a finite number of closed sets with diameter ≤ 1. One of these sets, which we call F_1, cannot be covered by finitely many elements of $\mathcal{U}$. So we can continue the process with F_1 instead of F. We get a decreasing sequence of closed sets $F_1, F_2, \ldots$ such that diam $F_n \leq 1/n$ and no F_n can be covered by finitely many elements of $\mathcal{U}$. We choose $x_n \in F_n$, and (x_n) is then a Cauchy sequence converging to $x \in \cap_{n=1}^{\infty} F_n$. But $x \in U$ for some $U \in \mathcal{U}$. Because diam F_n goes to zero and $x \in F_n$, we have $F_n \subset U$ for n sufficiently large which gives a contradiction. $\square$

THEOREM 2.2.4. *Let X be a Banach space with closed unit ball B. If $T \in \mathfrak{L}(X)$ is a limit of finite-rank operators then the closure of $T(B)$ is compact in X.*

PROOF. By Lemma 2.2.3 we have to prove that $\overline{T(B)}$ verifies property (ii). Let $\epsilon > 0$ be given. There exists $S \in \mathfrak{F}$ such that $\|T - S\| < \epsilon/3$. Because $\overline{S(B)}$ is compact there exist $x_1, \ldots, x_n \in B$ such that $S(B)$ is included in $B(x_1, \frac{\epsilon}{3}) \cup \cdots \cup B(x_n, \frac{\epsilon}{3})$. Consequently $T(B) \subset B(x_1, \frac{2\epsilon}{3}) \cup \cdots \cup B(x_n, \frac{2\epsilon}{3})$ and hence $\overline{T(B)} \subset B(x_1, \epsilon) \cup \cdots \cup B(x_n, \epsilon)$. $\square$

This property suggests the introduction of a new class of operators which are in some sense not far from finite-rank operators.

DEFINITION. Given a Banach space X we say that a linear operator T from X into X is *compact* if the closure of the image by T of the closed unit ball is compact in X.

This is equivalent to saying that for every bounded sequence (x_n) in X the sequence (Tx_n) contains a converging subsequence. This definition was introduced by F. Riesz in 1918 in order to generalize the results of I. Fredholm on integral equations (see Exercice II.4). As we shall see in Theorem 2.2.10, compact operators are spectrally very similar to finite-dimensional ones.

Let us denote by $\mathfrak{LC}(X)$ the set of compact operators on X. With the argument of Theorem 2.2.4 it is easy to see that $\mathfrak{LC}(X)$ is a closed two-sided ideal of $\mathfrak{L}(X)$ which, of course, contains $\overline{\mathfrak{F}}$. Is it true that $\overline{\mathfrak{F}} = \mathfrak{LC}(X)$? Even if this is true for many concrete Banach spaces (we shall give a proof of this approximation of compact operators by finite-rank operators on a Hilbert space in Corollary 2.3.5), it is false in general. This was first proved in 1973 by P. Enflo. Later his argument was greatly simplified by A.M. Davie. Inspite of this, D.C. Kleinecke proved in 1963 that $\overline{\mathfrak{F}}$ and $\mathfrak{LC}(X)$ are not far spectrally (see Corollary 5.7.3).

Let $T \in \mathfrak{L}(X)$. By definition the *spectrum* of T is the set of $\lambda \in \mathbb{C}$ such that $T - \lambda I$ is not invertible in $\mathfrak{L}(X)$. It is denoted by $\operatorname{Sp} T$. So $\lambda \in \operatorname{Sp} T$ if and only if at least one of the following conditions is verified:

(a) the range of $T - \lambda I$ is not all of X,

(b) $T - \lambda I$ is not injective.

We denote by $\mathcal{N}(T)$ the kernel of T and by $\mathcal{R}(T)$ its range.

If (b) holds we say that the λ is an *eigenvalue* of T. The corresponding *eigenspace* $\mathcal{N}(T - \lambda I)$ is the set of x such that $Tx = \lambda x$. Such an $x \neq 0$ is called an *eigenvector* corresponding to λ.

If $T \in \mathfrak{LC}(X)$ and $\dim X = \infty$ then $0 \in \operatorname{Sp} T$: otherwise T would be surjective, thus $I = T \cdot T^{-1}$ would be compact, and by Theorem 1.1.2 this would imply that X is finite-dimensional.

THEOREM 2.2.5. *Let X be a Banach space and let $T \in \mathfrak{LC}(X)$.*

(i) *If the range of T is closed, then it is finite-dimensional.*

(ii) *If $\lambda \neq 0$, then $\dim \mathcal{N}(T - \lambda I) < +\infty$.*

PROOF. (i) If $\mathcal{R}(T)$ is closed then it is complete so, by Theorem 1.1.14, T is an open mapping from X onto $\mathcal{R}(T)$. This implies that $\mathcal{R}(T)$ is locally compact so, by Theorem 1.1.2, $\mathcal{R}(T)$ is finite-dimensional. Let $Y = \mathcal{N}(T - \lambda I)$. The restriction of T to Y is a compact operator with range Y. So (ii) follows from (i). $\square$

If P is a compact projection, that is P is compact and satisfies $P^2 = P$, then $\mathcal{R}(P)$ is closed. This is because if Px_n converges to y then $P^2x_n = Px_n$ converges to Py, so $y = Py \in \mathcal{R}(P)$, and hence by Theorem 2.2.5 (i) $\mathcal{R}(P)$ is finite-dimensional.

By definition the adjoint $T^* \in \mathcal{L}(X')$ of an operator $T \in \mathcal{L}(X)$ satisfies $\langle Tx, f \rangle = \langle x, T^*f \rangle$, for all x in X and all f in the topological dual X'.

THEOREM 2.2.6.   *Let X be a Banach space. Then T is compact if and only if its adjoint T^* is compact.*

PROOF.   Suppose that T is compact. Let (f_n) be a sequence in the unit ball of X'. Because $|f_n(x) - f_n(y)| \leq \|x - y\|$ the family $\{f_n\}$ is equicontinuous. But $\overline{T(B)}$ is compact so, by Theorem 1.1.18, there exists a subsequence (f_{n_k}) converging uniformly on $T(B)$. But $\|T^*f_{n_i} - T^*f_{n_j}\| = \sup_{x \in B} |(f_{n_i} - f_{n_j})(Tx)|$, so the sequence $(T^*f_{n_i})$ converges. By the remark following the definition of compact operators this implies that T^* is compact.

Suppose now that T^* is compact. Let i be the canonical isometry of X into the bidual X'' defined by $i(x)(f) = f(x)$ for $x \in X$, $f \in X'$. Using the usual duality notations we have $\langle f, i(Tx) \rangle = \langle Tx, f \rangle = \langle x, T^*f \rangle = \langle T^*f, i(x) \rangle = \langle f, T^* \circ i(x) \rangle$ for all $x \in X$, $f \in X'$. So $i \circ T = T^* \circ i$. If $x \in B$ then $i(x) \in B''$, the unit ball of X'', so $i(T(B)) \subset T^{**}(B'')$. By the first part of the theorem, T^{**} is compact so $T^{**}(B'')$ has a compact closure. The same is true for $i(T(B))$. But i being an isometry, $T(B)$ has a compact closure. ◻

We now give F. Riesz's theory of compact operators.

LEMMA 2.2.7.   *Let X be a Banach space, let $T \in \mathcal{L}(X)$ and $\lambda \in \mathbb{C}$. Suppose that E, F are two closed subspaces of X such that $E \subset F$, $E \neq F$ and $(T - \lambda I)(F) \subset E$. Then there exists $a \in F \backslash E$ with $\|a\| \leq 1$, such that $\|Ta - Tx\| \geq |\lambda|/2$, for all $x \in E$.*

PROOF   By hypothesis, there exists $b \in F \backslash E$ such that $\operatorname{dist}(b, E) = \alpha > 0$. Let $y \in E$ be such that $\|b - y\| \leq 2\alpha$ and let $a = \frac{b-y}{\|b-y\|} \in F$. We have $\|a\| = 1$. For all $z \in E$, we have

$$a - z = \frac{1}{\|b - y\|}(b - y - z\|b - y\|) \quad \text{and} \quad y + z\|b - y\| \in E,$$

so $\big\|b - y - z\|b - y\|\big\| \geq \alpha$ and consequently $\|a - z\| \geq 1/2$ for all $z \in E$. But for $x \in E$ we have $T(a - x) = (T - \lambda I)(a - x) + \lambda(a - x)$ and $(T - \lambda I)(a - x) - \lambda x \in E$ by hypothesis, so $\|Ta - Tx\| \geq |\lambda|/2$. ◻

THEOREM 2.2.8. Let X be a Banach space and let $T \in \mathfrak{LC}(X)$. If $\lambda \neq 0$ then $T - \lambda I$ has a closed range and $\operatorname{codim} \mathcal{R}(T - \lambda I) < +\infty$.

PROOF. By Theorem 2.2.5 (ii), $\dim \mathcal{N}(T - \lambda I) < +\infty$. So using the argument given before Lemma 2.2.1, we see that there exists a closed subspace M such that $M + \mathcal{N}(T - \lambda I) = X$ and $M \cap \mathcal{N}(T - \lambda I) = \{0\}$. Let S be the restriction of $T - \lambda I$ to M with values in X. Then S is injective on M because $Sx = 0$ for $x \in M$ implies $x \in M \cap \mathcal{N}(T - \lambda I)$. Because $X = M + \mathcal{N}(T - \lambda I)$ we have $\mathcal{R}(S) = \mathcal{R}(T - \lambda I)$. We now prove that there exists $r > 0$ such that

$$r\|x\| \leq \|Sx\|, \quad \text{for } x \in M. \tag{1}$$

If this inequality is false then there exists a sequence (x_n) of elements of M such that $\|x_n\| = 1$ and $\lim_n Sx_n = 0$. But T is compact, so there exists a subsequence (x_{n_k}) such that Tx_{n_k} converges to some $a \in X$. Then λx_{n_k} converges to a, so $a \in M$ and $Sa = \lim(\lambda Sx_{n_k}) = 0$. Consequently $a = 0$, which is absurd because $\|x_n\| = 1$. Thus (1) is true. From (1) we conclude that $\mathcal{R}(S)$ is closed. For if Sx_n converges, then (Sx_n) is a Cauchy sequence so, by (1), (x_n) is a Cauchy sequence which converges to some $b \in M$ and hence $\lim_n Sx_n = Sb$. If $\mathcal{R}(T - \lambda I)$ has infinite codimension then there exists a sequence (a_n) in X such that, for all n, (a_n) is not in the subspace V_{n-1} generated by $\mathcal{R}(T - \lambda I)$ and $a_1, \ldots, a_{n-1}$. By the first part V_n is closed. By Lemma 2.2.7 there exists a sequence (b_n) such that $b_n \in V_n \backslash V_{n-1}$, $\|b_n\| \leq 1$ and $\|Tb_n - Tb_i\| \geq |\lambda|/2 > 0$, $1 \leq i \leq n - 1$. This implies that the sequence (Tb_n) has no converging subsequence which contradicts the fact that T is compact. $\square$

THEOREM 2.2.9. Let X be a Banach space and let $T \in \mathfrak{LC}(X)$. Suppose that $\lambda \neq 0$ and define $N_k = \mathcal{N}((T - \lambda I)^k)$ and $R_k = \mathcal{R}((T - \lambda I)^k)$ for $k \geq 1$. Then

(i) the N_k form an increasing sequence of finite-dimensional subspaces of X and the R_k form a decreasing sequence of closed subspaces of X having finite codimensions,

(ii) there exists a smallest integer n such that $N_{k+1} = N_k$ for $k \geq n$, and then $R_{k+1} = R_k$ for $k \geq n$, and we have

$$X = N_n \oplus R_n,$$

(iii) N_n and R_n are invariant subspaces of $T - \lambda I$, the restriction of $T - \lambda I$ to N_n has a nth power zero in $\mathfrak{L}(N_n)$ and the restriction to R_n is invertible in $\mathfrak{L}(R_n)$.

PROOF. It is obvious that $(T - \lambda I)^k = (-\lambda I)^k + K_k$ where K_k is a compact operator. So (i) derives immediately from Theorem 2.2.5 (ii) and Theorem 2.2.8.

We now prove (ii). Suppose that $N_k \neq N_{k+1}$ for all $k \geq 1$. Then we have $(T - \lambda I)(N_{k+1}) \subset N_k$ so, by Lemma 2.2.7, there exists a sequence (x_k) such that $x_k \in N_k \backslash N_{k-1}$, $\|x_k\| \leq 1$ for $k > 1$ and $\|Tx_k - Tx_i\| \geq |\lambda|/2$ for $i < k$. This gives a contradiction to the compactness of T. A similar argument also proves that the sequence (R_k) stabilizes. Let n be the smallest integer such that $N_k = N_{k+1}$, for $k \geq n$. We now prove that $N_n \cap R_n = \{0\}$. If $y \in N_n \cap R_n$ there exists $x \in X$ such that $y = (T - \lambda I)^n x$. Moreover $(T - \lambda I)^n y = 0$, and so $(T - \lambda I)^{2n} x = 0$. Therefore $x \in N_{2n} = N_n$, so $y = 0$. Because (R_k) stabilizes there exists a smallest $m \geq n$ such that $R_k = R_{k+1}$ for $k \geq m$. We have $R_m \subset R_n$ and $(T - \lambda I)R_m = R_m$. We now prove that $R_m = R_n$ so that $m = n$. If we have $m > n$, let $z \in R_{m-1} \subset R_n$ with $z \notin R_m$. Since $(T - \lambda I)z \in R_m = (T - \lambda I)R_m$ there exists $t \in R_m$ such that $(T - \lambda I)t = (T - \lambda I)z$. Hence $z - t \in N_1 \subset N_n$, but also $z - t \in R_n$, so $z = t$ and this is a contradiction. We now prove that $X = N_n + R_n$. For all $x \in X$ we have $(T - \lambda I)^n x \in R_n = R_m$ and $(T - \lambda I)^n R_n = R_n$, so there exists $y \in R_n$ such that $(T - \lambda I)^n y = (T - \lambda I)^n x$. Thus $x - y \in N_n$, hence the result.

We now prove (iii). If $x \in N_n$ then $(T - \lambda I)^{n+1} x = 0$ so $(T - \lambda I)x \in N_n$ and obviously $(T - \lambda I)^n = 0$ on N_n. If $x \in R_n$ then $x = (T - \lambda I)^n y$ for some $y \in X$, so $(T - \lambda I)x \in R_{n+1} = R_n$. By definition, the range of $T - \lambda I$ restricted to R_n is $R_{n+1} = R_n$ so this restriction is surjective. This restriction is also injective because $N_n \cap R_n = \{0\}$. $\square$

This theorem is particularly interesting when λ is an eigenvalue of T. For instance, it immediately implies that for $\lambda \neq 0$ the following properties are equivalent:

(i) $\mathcal{N}(T - \lambda I) = \{0\}$,

(ii) $\mathcal{R}(T - \lambda I) = X$,

(iii) λ is not in the spectrum of T.

In fact the decomposition of X as a direct sum of two closed subspaces E and F which are such that:

(a) $\dim E < +\infty$,

(b) $T(E) \subset E$ and the restriction of $T - \lambda I$ on E is nilpotent in $\mathcal{L}(E)$,

(c) $T(F) \subset F$ and the restriction of $T - \lambda I$ on F is invertible in $\mathcal{L}(F)$,

is unique.

Indeed, let $x \in E$, then $x = y + z$ where $y \in N_n$, $z \in R_n$. By hypothesis there exists an integer p such that $(T - \lambda I)^p x = 0$, so $(T - \lambda I)^p y = -(T - \lambda I)^p z \in$

$N_n \cap R_n = \{0\}$ and hence $(T - \lambda I)^p z = 0$. But the restriction of $T - \lambda I$ on R_n is invertible, so $z = 0$. Consequently $E \subset N_n$. A similar argument shows that $N_n \subset E$, hence $N_n = E$. If $x = y + z \in F$ with $y \in N_n$, $z \in R_n$ we have $(T - \lambda I)^n x = (T - \lambda I)^n z \in R_n$ so $(T - \lambda I)^n(F) \subset R_n$. But $(T - \lambda I)^n(F) = F$ by (c), so $F \subset R_n$. A similar argument proves that $R_n \subset F$.

THEOREM 2.2.10. *Let X be a Banach space and let $T \in \mathfrak{LC}(X)$. Then we have the following properties:*

 (i) *the spectrum of T is a compact subset of $\mathbb{C}$ having at most 0 as a limit point,*

(ii) *every spectral value $\lambda \neq 0$ is an eigenvalue of T.*

PROOF. The compactness of $\operatorname{Sp} T$ follows from Theorem 3.2.8, which we shall prove later (see also Exercise II.6). By the remark following Theorem 2.2.9, if $\lambda \neq 0$ is a spectral value then $\mathcal{N}(T - \lambda I) \neq 0$. So (ii) is proved. Let $\lambda_0 \in \operatorname{Sp} T$, $\lambda_0 \neq 0$ be an eigenvalue and let $E(\lambda_0) = N_n$, $F(\lambda_0) = R_n$ be the corresponding closed subspaces obtained in Theorem 2.2.9. We have $0 < \dim E(\lambda_0) < +\infty$. Let us denote by T_1 (resp. T_2) the restriction of T to $E(\lambda_0)$ (resp. $F(\lambda_0)$). Then $T_1 - \lambda_0 I_{E(\lambda_0)}$ is nilpotent on the finite-dimensional subspace $E(\lambda_0)$. So the characteristic polynomial of T_1 is $p(z) = (\lambda_0 - z)^p$ for some integer $p \leq n$. This implies in particular that $T_1 - \lambda I_{E(\lambda_0)}$ is invertible in $\mathfrak{L}(E(\lambda_0))$ for $\lambda \neq \lambda_0$. By Theorem 2.2.9 (iii), we have $T_2 - \lambda_0 I_{F(\lambda_0)}$ invertible in $\mathfrak{L}(F(\lambda_0))$ so, by Theorem 3.2.3 (which we will prove later), $T_2 - \lambda I_{F(\lambda_0)}$ is invertible in $\mathfrak{L}(F(\lambda_0))$ for $|\lambda - \lambda_0| < \epsilon$, with ϵ small enough. These two results imply that $\mathcal{N}(T - \lambda I) = \{0\}$ and $\mathcal{R}(T - \lambda I) = X$ for $0 < |\lambda - \lambda_0| < \epsilon$, so that $T - \lambda I$ is invertible for $0 < |\lambda - \lambda_0| < \epsilon$. Consequently λ_0 is isolated in the spectrum T, hence we have (i). $\square$

If a compact operator has a non-zero spectrum then it has an eigenvalue and it can be decomposed. If it has only 0 as a spectral value then we may have an eigenvalue at 0, but there are also examples without eigenvalue. We now give such an example.

EXAMPLE. Let $X = C([0,1])$ and let T be the *Volterra operator* defined by

$$(Tf)(x) = \int_0^x f(t)\, dt \ , \quad \text{for all } f \in C([0,1]).$$

We have $\|Tf\| \leq \|f\|$ and $\frac{d}{dx}(Tf)(x) = f(x)$. So if (f_n) is a sequence of the unit ball of X then (Tf_n) is equicontinuous and bounded so, by Theorem 1.1.18, (Tf_n) contains a converging subsequence. This shows that T is compact. If $\|f\| \leq 1$ it is obvious that $|Tf(x)| \leq x$ so by induction $|T^n f(x)| \leq x^n/n!$. This implies that

$\|T^n\| \le 1/n!$. So $\lim_{n\to\infty} \|T^n\|^{1/n} = 0$. Consequently $\operatorname{Sp} T = \{0\}$ by Theorem 3.2.8 (iii). Nevertheless 0 is not an eigenvalue, because $\int_0^x f(t)\,dt = 0$ for all x implies $f = 0$.

Given a Banach space X and a linear subspace Y of X, we say that Y is an *invariant subspace* of the bounded linear operator T defined on X if $T(Y) \subset Y$. Obviously $\{0\}$ and X are invariant subspaces and Y invariant implies $\overline{Y}$ invariant. So the interesting closed invariant subspaces are the non-trivial ones. Given an operator T, it is interesting to characterize all its closed invariant subspaces because in this way we can get more information about the structure and the spectrum of T.

In 1930, J. von Neumann proved that every compact operator on a Hilbert space has a non-trivial closed invariant subspace. This was extended in 1954 by N. Aronszajn and K.T. Smith for compact operators on a Banach space. Using non-standard analysis, A.R. Bernstein and A. Robinson proved in 1966 that the same result is true for a polynomially compact operator T, that is an operator for which there exists a polynomial p such that $p(T)$ is compact. Later on in 1966, P. Halmos obtained the same result using classical analysis. But all the arguments they used are complicated. In 1973, V.I. Lomonosov applied the Schauder Fixed Point Theorem and a nice argument on compactness to obtain a wonderful and simple proof of a theorem extending all the previous ones. H.M. Hilden even succeeded in eliminating Schauder's theorem.

We now present their arguments.

THEOREM 2.2.11 (V.I. LOMONOSOV). *Let K be a non-zero compact operator on an infinite-dimensional Banach space X. Then there exists a non-trivial closed linear subspace of X which is invariant under all bounded linear operators commuting with K.*

PROOF. Suppose that the assertion is false. If λ is an eigenvalue of K then we have $\mathcal{N}(K - \lambda I)$ invariant by all the operators commuting with K, so we get a contradiction. Consequently, K has no eigenvalues. So 0 is the only spectral value of K. Without loss of generality we may suppose that $\|K\| = 1$. Let $x_0 \in X$ be such that $\|Kx_0\| > 1$, then we have $\|x_0\| = \|K\|\,\|x_0\| > 1$. Let S be the closed unit ball with centre at x_0. Because $\|x_0\| > 1$ we have $0 \notin S$.

Moreover for all $x \in S$ we have $\|Kx - Kx_0\| \le \|x - x_0\| \le 1$ so $\overline{K(S)}$ is included in the closed unit ball with centre at Kx_0. Because $\|0 - Kx_0\| = \|Kx_0\| > 1$ we conclude that $0 \notin \overline{K(S)}$. Let A be the closed subalgebra in $\mathcal{L}(X)$ of all T such that $KT = TK$. For all $y \ne 0$, $Y = \{Ty : T \in A\}$ is then an invariant subspace for all $T \in A$ and $Y \neq \{0\}$, so it is dense in X. Consequently, for all $y \neq 0$ there exists $T \in A$ such that $\|Ty - x_0\| < 1$. For $T \in A$, we consider the open set $\omega(T) = \{y : \|Ty - x_0\| < 1\}$. Because K is a compact operator, the set $\overline{K(S)}$ is compact. All the $\omega(T)$ cover $X \backslash \{0\}$ and $0 \notin K(S)$, so there exists a finite subset $\mathfrak{F}$ of A such that $K(S)$ is covered by the $\omega(T)$ with $T \in \mathfrak{F}$. Consequently if $y \in K(S)$ there exists $T \in \mathfrak{F}$ such that $\|Ty - x_0\| < 1$. Until now this is the argument of Lomonosov. We now follow Hilden's argument. We have $Kx_0 \in K(S)$ so there exists $T_1 \in \mathfrak{F}$ such that $\|T_1 K x_0 - x_0\| < 1$, hence $T_1 K x_0 \in S$, and consequently $KT_1 K x_0 \in K(S)$. Thus there exists $T_2 \in \mathfrak{F}$ such that $\|T_2 K T_1 K x_0 - x_0\| < 1$, so $T_2 K T_1 K x_0 \in S$. By induction we can define a sequence (T_n) in $\mathfrak{F}$ such that

$$\|T_n K T_{n-1} \cdots T_2 K T_1 K x_0 - x_0\| < 1.$$

Let $\alpha = \max\{\|T\| : T \in \mathfrak{F}\}$. By induction it is easy to prove that we have $T_n K T_{n-1} \cdots T_2 K T_1 K = (T_n \cdots T_1) K^n$, so we conclude that:

$$\|T_n K T_{n-1} \cdots T_2 K T_1 K\| \leq \|(\alpha K)^n\|.$$

But $\operatorname{Sp} K = \{0\}$ implies $\operatorname{Sp}(\alpha K) = \{0\}$, so $\lim_{n \to \infty} \|(\alpha K)^n\| = 0$ by Theorem 3.2.8 (iii). Consequently $\lim_{n \to \infty} \|T_n K T_{n-1} \cdots T_2 K T_1 K x_0\| = 0$, so $\|x_0\| \leq 1$ which is a contradiction. $\square$

COROLLARY 2.2.12.   *Let T be a polynomially compact operator on an infinite-dimensional Banach space X. Then T has a non-trivial closed invariant subspace.*

PROOF.   Suppose that $p(T) = \alpha_0 I + \alpha_1 T + \cdots + \alpha_n T^n$ is a compact operator. If $p(T) = 0$ let $x \neq 0$ in X. Then the finite-dimensional linear subspace generated by $x, Tx, \ldots, T^n x$ is non-trivial and invariant by T. If $p(T) \neq 0$ then T commutes with $p(T)$, and we then apply Theorem 2.2.11. $\square$

Is it true that every bounded linear operator on a Banach space has a non-trivial closed invariant subspace? In 1975, P. Enflo gave a counter-example to that conjecture, but his very long preprint contained many gaps (which can be removed in the opinion of C. Beauzamy). Finally, a complete counter-example was given by C. Read in 1984, and A.M. Davie subsequently produced an even easier one. But the problem is still open for operators on a Hilbert space.

If X is an arbitrary Banach space we have been able to build non-trivial elements in the subalgebra $\mathbb{C}I + \mathfrak{LC}(X)$. Is it possible to exhibit a bounded linear operator which is not in this subalgebra? Of course, if X is a concrete Banach space, it is possible. But if X is an arbitrary Banach space it is still an open problem to know if $\mathbb{C}I + \mathfrak{LC}(X)$ is different from $\mathfrak{L}(X)$. The most recent result towards the solution of this problem is given in S. Shelah and J. Steprans, *A Banach space on which there are few operators*, Proc. Amer. Math. Soc. **104** (1988), pp. 101–105.

§3. Bounded Linear Operators on a Hilbert Space

There are many examples of bounded linear operators on a Hilbert space. For instance, if E is a closed linear subspace of H and if $E^\perp$ denotes its orthogonal complement then we have $H = E \oplus E^\perp$. So for every $x \in H$, there is a unique decomposition $x = x_1 + x_2$ such that $x_1 \in E$, $x_2 \in E^\perp$. The linear mapping P from H onto E defined by $Px = x_1$ is called the *orthogonal projection* on E. It satisfies $P = P^2$, $\|P\| = 1$, and obviously $I - P$ is the orthogonal projection on $E^\perp$ and satifies $I - P = (I - P)^2$ and $\|I - P\| = 1$.

Let $(E_i)_{i \in I}$ be a family of orthogonal closed subspaces of H such that H is the Hilbertian direct sum of the E_i and let $(\lambda_i)_{i \in I}$ be a bounded family of complex numbers. Denoting by P_i the orthogonal projection on E_i, it is easy to verify that for a fixed $x \in H$ the family

$$T_F x = \sum_{i \in F} \lambda_i P_i x$$

is a Cauchy family, for all finite subsets F of I. Let $\epsilon > 0$. There exists a finite subset F_0 of I such that $\sum_{i \in F} \|P_i x\|^2 \leq \epsilon$ for all finite subsets F of I disjoint from F_0. Consequently, if we take finite subsets F_1 and F_2 of I such that $F_0 \subset F_1 \subset F_2$ we have $\|T_{F_1} x - T F_2 x\|^2 = \|\sum_{i \in F_2 \setminus F_1} \lambda_i P_i x\|^2 = \sum_{i \in F_2 \setminus F_1} |\lambda_i|^2 \|P_i x\|^2 \leq C^2 \sum_{i \in F_2 \setminus F_1} \|P_i x\|^2 \leq C^2 \epsilon$, where $\sup |\lambda_i| \leq C$. So this implies that the family $(T_F x)$ converges in H. If we denote by T the linear operator defined by

$$T x = \sum_{i \in I} \lambda_i P_i x$$

then this operator is bounded by Theorem 1.1.13. We call it the *strong sum* of the series $\sum_{i \in I} \lambda_i P_i$. It is important to notice that in general such a series converges strongly but *not in norm*.

If the family $(\lambda_i)_{i \in I}$ goes to 0, that is for every $\epsilon > 0$ there exists a finite subset $F_0 \subset I$ such that $\max_{i \in F} |\lambda_i| \leq \epsilon$, for all finite subsets $F \subset I$ disjoint from F_0, then the previous series converges in norm. As previously, for an arbitrary x we have

$$\|T_{F_1} x - T_{F_2} x\|^2 = \sum_{i \in F_2 \setminus F_1} |\lambda_i|^2 \|P_i x\|^2 \leq \epsilon^2 \sum_{i \in F_2 \setminus F_1} \|P_i x\|^2 \leq \epsilon^2 \|x\|^2,$$

so $\|T_{F_1} - T_{F_2}\| \leq \epsilon$. This implies that (T_F) converges in norm in $\mathcal{L}(H)$ to some T and we have

$$T = \sum_{i \in I} \lambda_i P_i.$$

This situation will occur for self-adjoint and normal compact operators on a Hilbert space (see below).

Another interesting class of bounded linear operators on a Hilbert space is given by the *weighted shifts*. Suppose that H is a separable Hilbert space having an orthonormal basis $e_1, e_2, \ldots$. Then for all $x = \sum_{n=1}^{\infty} \lambda_n e_n$ with $\sum_{n=1}^{\infty} |\lambda_n|^2 < +\infty$ we define the right shift with weight given by a bounded sequence (α_n) by

$$Tx = \sum_{n=1}^{\infty} \alpha_n \lambda_n e_{n+1}.$$

It is easy to verify that $\|T\| \leq \max_n |\alpha_n|$. We can also define the left shift with weight given by a bounded sequence (α_n) by

$$Sx = \sum_{n=1}^{\infty} \alpha_n \lambda_n e_{n-1} \quad \text{(with the convention that } e_0 = 0).$$

It is also easy to verify that $\|S\| \leq \max_n |\alpha_n|$. Of course T and S are not invertible because $e_1 \notin \mathcal{R}(T)$ and $e_1 \in \mathcal{N}(S)$. Suppose that the weight is one, that is $\alpha_n = 1$ for all n. Then we have $ST = I$ and $TS =$ projection on the orthogonal complement of $e_1 \neq I$.

For more concrete examples of bounded linear operators on a Hilbert space the reader is refered to the classical book by P. Halmos [3].

If $T \in \mathcal{L}(H)$ then the *adjoint of T* denoted by T^* satisfies the property $(Tx|y) = (x|T^*y)$, for all $x, y \in H$.

Then $T \mapsto T^*$ is an *involution* on $\mathcal{L}(H)$, that is it satisfies the following properties:

(a) $(T + S)^* = T^* + S^*$,

(b) $(\lambda T)^* = \overline{\lambda} T^*$,

(c) $(T^*)^* = T$,

(d) $(TS)^* = S^* T^*$.

THEOREM 2.3.1. *Let H be a Hilbert space and let $T \in \mathcal{L}(H)$. Then we have the following properties:*

(i) $\mathcal{N}(T) = \mathcal{R}(T^*)^{\perp}$ and $\mathcal{N}(T^*) = \mathcal{R}(T)^{\perp}$,

(ii) $\|T\| = \|T^*\|$,

(iii) $\|T^*T\| = \|T\|^2$,

(iv) $I + T^*T$ is invertible in $\mathfrak{L}(H)$ and $\|(I + T^*T)^{-1}\| \leq 1$.

PROOF.   Obviously $x \in \mathcal{N}(T)$ is equivalent to $(Tx|y) = (x|T^*y) = 0$ for all $y \in H$, so $x \in \mathcal{R}(T^*)^\perp$. The second relation comes from the first one and (c). We have $\|Tx\|^2 = (Tx|Tx) = (T^*Tx|x) \leq \|T^*T\|\,\|x\|^2 \leq \|T^*\|\,\|T\|\,\|x\|^2$. Consequently $\|T\|^2 \leq \|T^*T\| \leq \|T^*\|\,\|T\|$. This implies that $\|T\| \leq \|T^*\|$. So we have (ii) by (c) and simultaneously we have property (iii). Moreover

$$\|(I + T^*T)x\|^2 = ((I + T^*T)^2 x|x) = \|x\|^2 + 2\|Tx\|^2 + \|T^*Tx\|^2 \geq \|x\|^2. \qquad (1)$$

So $I + T^*T$ is injective. The same inequality implies that its range is closed. If $((I + T^*T)x_n)$ is a Cauchy sequence then (x_n) is also a Cauchy sequence. By (i) we have $\mathcal{R}(I + T^*T)^\perp = \mathcal{N}(I + T^*T) = \{0\}$ so $\mathcal{R}(I + T^*T) = H$, hence $I + T^*T$ is invertible in $\mathfrak{L}(H)$. Moreover by inequality (1) we have $\|(I + T^*T)^{-1}\| \leq 1$. $\square$

Among bounded linear operators three classes of operators are interesting.

DEFINITION.   An operator $T \in \mathfrak{L}(H)$ is said to be

(i) *self-adjoint* if $T = T^*$,

(ii) *normal* if $TT^* = T^*T$,

(iii) *unitary* if $TT^* = T^*T = I$.

Every $T \in \mathfrak{L}(H)$ has a unique decomposition $T = R + iS$ where $R, S \in \mathfrak{L}(H)$ and $R = R^*$, $S = S^*$. These, the *real part* and *imaginary part* of T, are given by $R = (T + T^*)/2$, $S = (T - T^*)/2i$. Also it is easy to prove that T is unitary if and only if T is an isometry of H onto itself.

For $T \in \mathfrak{L}(H)$ we denote by $\rho(T)$ the *spectral radius* of T, that is $\rho(T) = \max\{|\lambda|: \lambda \in \operatorname{Sp} T\}$.

COROLLARY 2.3.2.   *Let H be a Hilbert space and let S be a self-adjoint operator on H. Then we have the following properties:*

(i) $\|S^2\| = \|S\|^2$ *and consequently* $\|S\| = \rho(S)$,

(ii) $\operatorname{Sp} S \subset \mathbf{R}$ *and consequently* $\operatorname{Sp}(T^*T) \subset \mathbf{R}_+$ *for all* $T \in \mathfrak{L}(H)$.

PROOF.   By Theorem 2.3.1 (iii) we have $\|S^2\| = \|S\|^2$. Consequently by induction we get $\|S^{2^n}\| = \|S\|^{2^n}$. By Theorem 3.2.8, which we shall prove later, we have

$\rho(S) = \lim_{n \to \infty} \|S^n\|^{1/n}$, so $\|S\| = \rho(S)$. Let $\lambda = \alpha + i\beta$ with $\alpha, \beta \in \mathbf{R}$. Denote $\lambda I - S$ by S_λ. Then $S_\lambda = S_\alpha + i\beta I$. So for all $x \in H$ we have

$$\|S_\lambda x\|^2 = (S_\lambda x | S_\lambda x) = (S_\alpha x | S_\alpha x) + |\beta|^2 \|x\|^2$$

and consequently

$$\|S_\lambda x\| \geq \beta \|x\|.$$

As in the proof of Theorem 2.3.1 we conclude that $S_\lambda = \lambda I - S$ is invertible for $\beta \neq 0$, so $\mathrm{Sp}\, S \subset \mathbf{R}$. By Theorem 2.3.1 (iv), $I + \lambda^2 T^\star T$ is invertible for all $\lambda > 0$ (we apply it to λT) so $\mathrm{Sp}(T^\star T)$ cannot contain a negative number. $\square$

It is easy to prove that the algebra $\mathfrak{LC}(H)$ of compact operators is stable by the involution $T \mapsto T^\star$. The argument is very similar to that of Theorem 2.2.6.

We now study the particular case of self-adjoint compact operators on a Hilbert space which has several consequences for integral and differential equations.

THEOREM 2.3.3. *Let H be a Hilbert space and let $T \in \mathfrak{LC}(H)$ be self-adjoint. Then either $\|T\|$ or $-\|T\|$ is an eigenvalue of T.*

PROOF. If $T = 0$ the result is obvious. So suppppose that $T \neq 0$. By Corollary 2.3.2 we have $\mathrm{Sp}\, T \subset \mathbf{R}$. Moreover, by Corollary 2.3.2, we have $\|T\| = \rho(T)$ so there exists $\lambda \neq 0$, $\lambda \in \mathrm{Sp}\, T$ such that $|\lambda| = \|T\|$. Consequently $\lambda = \|T\|$ or $\lambda = -\|T\|$ and by Theorem 2.2.10 it is an eigenvalue. $\square$

We know that self-adjoint $n \times n$ matrices can be diagonalized or equivalently can be written as $\sum_{\alpha=1}^{k} \lambda_\alpha P_\alpha$ where the P_α are self-adjoint orthogonal projections and the λ_α are real numbers. This result can be extended to self-adjoint compact operators on a Hilbert space by the following:

THEOREM 2.3.4 (SPECTRAL THEOREM FOR SELF-ADJOINT COMPACT OPERA-TORS ON A HILBERT SPACE). *Let H be a Hilbert space and let T be a self-adjoint compact operator on H. Let $\{\lambda_k\}_{k \geq 1}$ be the discrete set of non-zero eigenvalues of T. Also let $E_0 = \mathcal{N}(T)$ and $E_k = \mathcal{N}(T - \lambda_k I)$, for $k \geq 1$. Then we have the following properties:*

(i) *for $k \geq 0$ the closed subspaces E_k are orthogonal and their Hilbertian direct sum is H. Moreover if P_k denotes the self-adjoint projection on E_k we have $T P_k = P_k T$ for all k,*

(ii) the series $\sum_{k\geq 1} \lambda_k P_k$ converges in norm in $\mathfrak{L}(H)$ and we have

$$T = \sum_{k\geq 1} \lambda_k P_k.$$

PROOF. By Theorem 2.3.3, the set of non-zero eigenvalues is non-empty. So, by Theorem 2.2.10, it is either finite or a sequence converging to zero. If $x \in E_i$ and $y \in E_j$ with $i \neq j$ then we have $Tx = \lambda_i x$ and $Ty = \lambda_j y$ (with the convention that $\lambda_0 = 0$). So we have

$$\lambda_i(x|y) = (Tx|y) = \overline{\lambda}_j(x|y) = \lambda_j(x|y),$$

and consequently $(x|y) = 0$. Let $E = \overline{\oplus}_{k\geq 1} E_k$ be the Hilbertian direct sum of the E_k, that is the closure of the algebraic sum of the E_k. Because $T(E_k) = E_k$ and because T is bounded we have $T(E) \subset E$. The operator T being self-adjoint we also have $T(E^\perp) \subset E^\perp$. If T were different from zero on $E^\perp$ then its restriction to the closed subspace $E^\perp$ would be a self-adjoint compact operator so, by Theorem 2.3.3, it would have a non-zero eigenvalue which would also be a non-zero eigenvalue of T. But this is a contradiction for otherwise we would have $E^\perp \cap E_k \neq \{0\}$ for some $k \geq 1$. So $E^\perp = E_0$. Consequently $H = \overline{\oplus}_{k\geq 0} E_k$. If $\mathrm{Sp}\, T$ is finite it is obvious that $\sum_{k\geq 1} \lambda_k P_k$ converges. So suppose that $\mathrm{Sp}\, T$ is infinite.

Let $T_n = \sum_{1\leq k\leq n} \lambda_k P_k$. We have

$$\|T_{n+m}x - T_n x\|^2 = \|\sum_{k=n+1}^{n+m} \lambda_k P_k x\|^2 = \sum_{k=n+1}^{n+m} |\lambda_k|^2 \|P_k x\|^2$$

and $\sum_{k=n+1}^{n+m} \|P_k x\|^2 \leq \|x\|^2$ because the projections are orthogonal. Let $\epsilon > 0$. There exists a n_0 such that $k \geq n_0$ implies $|\lambda_k| \leq \epsilon$. So for $n \geq n_0$ we have $\|T_{n+m} - T_n\| \leq \epsilon$. Consequently the sequence (T_n) converges in $\mathfrak{L}(H)$ to some operator S. If $x \in E_p$ then $T_n x = \lambda_p x$ for all $n \geq p$, so $Sx = \lambda_p x = Tx$. Consequently, S and T coincide on all E_p for $p \geq 0$, so on their Hilbertian direct sum H we have $S = T$. $\square$

COROLLARY 2.3.5. *Let H be a Hilbert space. Then every compact operator on H can be approximated by finite-rank operators.*

PROOF. Let $T \in \mathfrak{LC}(H)$ and $\epsilon > 0$. Then we also have $T^* \in \mathfrak{LC}(H)$, so $\mathrm{Re}\, T = (T + T^*)/2$ and $\mathrm{Im}\, T = (T - T^*)/2i$ are self-adjoint compact operators on H. By Theorem 2.2.5 we know that $E_k = P_k(H)$ is finite dimensional. So, by Theorem 2.3.4 (ii), there exist two finite-rank operators T_1 and T_2 such that $\| \mathrm{Re}\, T - T_1\| < \epsilon/2$ and $\| \mathrm{Im}\, T - T_2\| < \epsilon/2$. Then $T_1 + iT_2$ has finite rank and $\|T - (T_1 + iT_2)\| < \epsilon$. $\square$

Theorem 2.3.4 can be reformulated differently.

COROLLARY 2.3.6 (FREDHOLM ALTERNATIVE). *Let H be an infinite-dimensional Hilbert space and let T be a self-adjoint compact operator on H. Denoting by $\{\lambda_k\}_{k \geq 1}$ the set of non-zero eigenvalues of T (of course $\lambda_k \in \mathbb{R}$), then we have the following properties:*

(i) *if $\lambda \neq \lambda_k$, for all k, then the equation $Tx - \lambda x = y$ has a unique solution in H for all $y \in H$,*

(ii) *if $\lambda \neq 0$ and $\lambda = \lambda_k$, for some k, then the equation $Tx - \lambda x = y$ has a solution in H if and only if y is orthogonal to $\mathcal{N}(T - \lambda I)$; so either this equation has no solution or it has an infinite number of solutions.*

All this theory can be applied to integral operators $Tf(x) = \int_a^b k(x, y) f(y) \, dy$ with symmetric kernels. In particular it has many applications to the Sturm-Liouville problem (see J. Dieudonné, *Eléments d'analyse*, Vol 1, Paris, 1963, Chapter 11, §7).

EXERCISE 1. Prove that a commutative and semi-simple finite-dimensional complex algebra is isomorphic to $\mathbf{C}^n$, where n denotes its dimension. What happens if it is a commutative and semi-simple finite-dimensional real algebra?

EXERCISE 2. Let f be an analytic function from a domain $D \subset \mathbf{C}$ into $M_n(\mathbf{C})$. Denote by σ_k the kth-symmetric function of the eigenvalues of a $n \times n$ matrix. Prove that $\lambda \mapsto \sigma_k(f(\lambda))$ is holomorphic on D.

∗EXERCISE 3. Prove that the commutators $[a, b] = ab - ba$ form a linear subspace of codimension one in $M_n(\mathbf{C})$ (Shoda's theorem). Given a linear functional f on $M_n(\mathbf{C})$ prove the equivalence of the following properties:

(i) $f = \alpha \cdot \text{Tr}$, where α is some complex number and where Tr denotes the trace,

(ii) $f(ab - ba) = 0$, for all $a, b \in M_n(\mathbf{C})$,

(iii) $f(xax^{-1}) = f(a)$, for all $a \in M_n(\mathbf{C})$ and x invertible in $M_n(\mathbf{C})$,

(iv) $|f(x)| \leq C\rho(x)$, where C is some positive constant and where ρ denotes the spectral radius.

EXERCISE 4.

(i) Suppose that k is continuous on $[a, b] \times [a, b]$. For $f \in C([a, b])$ define Tf by:

$$Tf(x) = \int_a^b k(x, y)f(y)\, dy.$$

Prove that T is a compact operator on $C([a, b])$.

(ii) Suppose that $k \in L^2([a, b] \times [a, b])$. For $f \in L^2([a, b])$ define Tf as previously. Prove that T is a compact operator on $L^2([a, b])$.

EXERCISE 5. Given a sequence (α_n) converging to 0, prove that there exists a compact operator T on ℓ^2 such that $\text{Sp}\, T = \{0\} \cup \{\alpha_n \mid n \in \mathbf{N}\}$.

EXERCISE 6. Given $T \in \mathcal{LC}(X)$ and $\epsilon > 0$ prove directly, without using Theorem 3.2.8, that $\{\lambda : \lambda \in \text{Sp}\, T,\ |\lambda| \geq \epsilon\}$ is finite.

EXERCISE 7. Given $T \in \mathcal{LC}(X)$ and $\lambda \neq 0$ prove that $\dim \mathcal{N}(T - \lambda I) = \text{codim}\, \mathcal{R}(T - \lambda I) = \dim \mathcal{N}(T^* - \overline{\lambda}I) = \text{codim}\, \mathcal{R}(T^* - \overline{\lambda}I)$.

EXERCISE 8. Let H be a separable Hilbert space with orthonormal basis $e_1, e_2, \ldots$. We consider T the left weighted shift having weight $\alpha_n = 1/n$ (resp. S the right weighted shift). Prove that T and S are compact operators on H, that T is nilpotent and that S is not nilpotent. Moreover show that $\operatorname{Sp} S = \{0\}$.

EXERCISE 9.   Let H be a Hilbert space and let I be a closed two-sided ideal of $\mathcal{L}(H)$. Prove that $I = \mathcal{LC}(H)$ (Calkin's theorem).

EXERCISE 10.   Let $H = L^2(]0, +\infty[)$. For $f \in H$ we define Tf by

$$(Tf)(x) = \frac{1}{x} \int_0^x f(t)\, dt.$$

Prove that $T \in \mathcal{L}(H)$ but that T is not compact.

*EXERCISE 11.   Given a right or left weighted shift T with weight (α_n) determine explicitly the spectrum of T. If you do not succeed look at [3].

EXERCISE 12.   Extend Theorem 2.3.4 for normal compact operators on a Hilbert space.

*EXERCISE 13.   Given $q, f \in C([a, b])$ and $\lambda \in \mathbf{C}$, the Sturm-Liouville problem consists in finding the solutions of the differential equation

$$y" - qy + \lambda y = f,$$

with some limit conditions $\alpha_1 y(a) + \beta_1 y'(a) = 0$, $\alpha_2 y(b) + \beta_2 y'(b) = 0$. Prove that y is a solution of this problem if and only if

$$y(x) = -\int_a^b K(t, x) f(t)\, dt$$

where K is a convenient continuous function on $[a, b] \times [a, b]$. If you do not succeed look at the book of J. Dieudonné.

Chapter III

BANACH ALGEBRAS

§1. Definition and Examples

A complex algebra is a vector space A over the complex field $\mathbb{C}$, with a multiplication satisfying the following properties:

$$x(yz) = (xy)z, \; (x+y)z = xz + yz, \; x(y+z) = xy + xz, \; \lambda(xy) = (\lambda x)y = x(\lambda y),$$

for all $x, y, z \in A$ and $\lambda \in \mathbb{C}$.

If moreover A is a Banach space for a norm $\|\cdot\|$ and satisfies the norm inequality $\|xy\| \leq \|x\| \cdot \|y\|$, for all $x, y \in A$, we say that A is a complex *Banach algebra*.

If A has a unit (which is unique), denoted by $\mathbf{1}$, we can always suppose that $\|\mathbf{1}\| = 1$ because otherwise we can replace $\|\cdot\|$ by an equivalent norm $\||\cdot\||$ satisfying $\||\mathbf{1}\|| = \mathbf{1}$ and $\||xy\|| \leq \||x\|| \cdot \||y\||$ for all $x, y \in A$ (see Exercise III.1). If A is a Banach algebra without unit it is always possible to imbed it isometrically in the Banach algebra with unit $\tilde{A} = A \times \mathbb{C}$, where the operations and the norm are defined by $(x, \alpha) + (y, \beta) = (x + y, \alpha + \beta)$, $\lambda(x, \alpha) = (\lambda x, \lambda \alpha)$, $(x, \alpha) \cdot (y, \beta) = (xy + \beta x + \alpha y, \alpha \beta)$, $\|(x, \alpha)\| = \|x\| + |\alpha|$. In the definition of a Banach algebra, instead of the condition $\|xy\| \leq \|x\| \cdot \|y\|$ it is enough to suppose that the left multiplications $x \to xy$ and right multiplications $x \to yx$ are continuous, because with this hypothesis there exists an equivalent norm satisfying the standard norm inequality (see Exercise III.2).

The reader may be surprised by the fact that we consider only Banach algebras over the field of complex numbers and not Banach algebras over the field of real numbers. The reason is simple. Very often the proofs involve the use of analytic tools which are inefficient in the case of the field of real numbers. In many cases, it is possible to extend to real algebras results obtained for complex algebras. However, the way to do this is not always obvious. So in this small book we shall restrict ourselves to the complex situation and, except otherwise stated, to the case of an algebra with unit. For the real situation, see the standard textbooks ([1], [2] and [6]).

We now give some examples of commutative Banach algebras.

EXAMPLE 1. Let K be a compact set. Then $C(K)$, the Banach space of all complex continuous functions on K, with the supremum norm, is a Banach algebra with unit. If K has only n elements then $C(K) = \mathbf{C}^n$ with coordinatewise multiplication, and of course $\mathbf{C}$ is the simplest commutative Banach algebra. If K is a locally compact space, then the Banach space of all complex continuous functions on K which go to zero at infinity, that is continuous functions f such that $\{x: x \in K, |f(x)| \geq \epsilon\}$ is compact for all $\epsilon > 0$, with the supremum norm, is a Banach algebra which has no unit if K is not compact.

EXAMPLE 2. If K is compact, every closed subalgebra of $C(K)$ is a Banach algebra. For instance if K is a compact subset of $\mathbf{C}^n$ the following examples of Banach algebras are interesting: $P(K)$, the algebra of continuous functions on K which are uniform limits of polynomials on K; $R(K)$, the algebra of continuous functions K which are uniform limits of rational functions with poles outside of K; and $A(K)$, the algebra of continuous functions on K which are holomorphic on the interior of K. We have the inclusions $P(K) \subset R(K) \subset A(K) \subset C(K)$, and there are examples in $\mathbf{C}^n$ (with $n \geq 2$) for which all these inclusions are strict.

EXAMPLE 3. Let G be a locally compact commutative group and let μ be its Haar measure. Then $L^1(G)$ is a Banach algebra if we define multiplication by convolution

$$(f * g)(x) = \int_G f(xy^{-1})g(y)\, d\mu(y)$$

and the norm by the L^1-norm $\|f\|_1 = \int_G |f(x)|\, d\mu(x)$. For instance if we consider the additive group $L^1(\mathbf{Z}) = l^1$ then we have

$$(f * g)(n) = \sum_{k \in \mathbf{Z}} f(n - k)g(k)$$

$$\|f\|_1 = \sum_{k \in \mathbf{Z}} |f(k)|.$$

In fact this algebra l^1 can be identified with the *Wiener algebra* W of continuous functions on $[0, 2\pi]$ having Fourier coefficients a_n such that $\sum_{n \in \mathbf{Z}} |a_n| < +\infty$

and $f(t) = \sum_{n\in\mathbf{Z}} a_n e^{int}$ on $[0, 2\pi]$, with the operations of pointwise addition and multiplication and the norm $\|f\| = \sum_{k\in\mathbf{Z}} |a_n|$.

EXAMPLE 4.   The Banach space of integrable functions on $[0, 1]$ with multiplication defined by convolution $(f*g)(x) = \int_0^x f(x-t)g(t)\,dt$ and L^1-norm $\|f\|_1 = \int_0^1 |f(t)|\,dt$ is a Banach algebra.

Let us now give examples of non-commutative Banach algebras.

EXAMPLE 5.   Let X be a Banach space. Then $\mathcal{L}(X)$, the algebra of all bounded linear operators on X, is a Banach algebra with unit for the usual operator norm. If X is finite dimensional, with dim $X = n$, then $\mathcal{L}(X)$ can be identified with $M_n(\mathbf{C})$. If dim $X > 1$ then $\mathcal{L}(X)$ is not commutative. Except for finite-dimensional ones, the simplest non-commutative Banach algebra is $\mathcal{L}(H)$ where H is an infinite-dimensional Hilbert space. This algebra $\mathcal{L}(H)$ has nice properties which we shall study in Chapter VI. Every closed subalgebra of $\mathcal{L}(X)$ is also a Banach algebra. For instance, the algebra of compact operators $\mathcal{LC}(X)$ is a Banach algebra which, by Theorem 1.1.2, has no unit if X is infinite-dimensional. Using Exercise III.3, it is easy to prove that every Banach algebra can be isomorphically represented as a closed subalgebra of $\mathcal{L}(X)$ for some Banach space X, but in practice this does not help very much.

Starting with some Banach algebras, how can new ones be obtained?

For instance, given A a Banach algebra, it is possible to consider $\mathcal{L}(A)$, the algebra of bounded linear operators on A, and all its closed subalgebras. Given a family $(A_i)_{i\in I}$ of Banach algebras we can define the *Banach algebra product* $\prod_{i\in I} A_i$ as the set of all families $(x_i)_{i\in I}$ such that $x_i \in A_i$ and $\sup_{i\in I} \|x_i\| < +\infty$ with the norm $\|x\| = \sup_{i\in I} \|x_i\|$. It is easy to verify that $\prod_{i\in I} A_i$ is a Banach algebra which contains an isometric copy of each A_{i_0} (by the mapping $x \to (x_i)_{i\in I}$ where $x_i = x$ if $i = i_0$ and $x_i = 0$ otherwise). If $I = \{1, \ldots, n\}$, obviously this product coincides with $A_1 \times \ldots \times A_n$.

Given a Banach algebra A it is also possible to consider the $n \times n$ *matrix algebra* $M_n(A)$ with the standard operations

$$(a_{ij}) + (b_{ij}) = (a_{ij} + b_{ij}), \quad (a_{ij}) \cdot (b_{ij}) = \left(\sum_{k=1}^n a_{ik} b_{kj}\right)$$

and the norm $\|(a_{ij})\| = \sup_{i=1,\ldots,n}(\|a_{i1}\| + \ldots + \|a_{in}\|)$.

$M_n(A)$ is not commutative for $n > 1$ and it contains an isometric copy of A (associating to x the diagonal matrix having only x on the diagonal).

A much more useful tool is the following. Let I be a closed two-sided ideal of A. Then A/I is a Banach space for the norm $|||\dot{x}||| = \inf_{u \in I} \|x + u\|$, where $\dot{x}$ denotes the coset $x + I$ of x. With this norm it is easy to verify that A/I is a Banach algebra, called the *quotient algebra* of A by the two-sided ideal I. This comes from

$$|||\dot{x} + \dot{y}||| \leq \|x + u + y + v\| \leq \|x + u\| + \|y + v\|$$

$$|||\dot{x} \cdot \dot{y}||| \leq \|xy + xv + uy + uv\| \leq \|x + u\| \cdot \|y + v\|$$

for $u, v \in I$, so $|||\dot{x} + \dot{y}||| \leq |||\dot{x}||| + |||\dot{y}|||$, $|||\dot{x} \cdot \dot{y}||| \leq |||\dot{x}||| \cdot |||\dot{y}|||$. This implies in particular that $|||\dot{1}^2||| = |||\dot{1}||| \leq |||\dot{1}||| \cdot |||\dot{1}|||$ so $1 \leq |||\dot{1}||| \leq \|1\| = 1$, and consequently $|||\dot{1}||| = 1$ if A has a unit.

EXAMPLE 6. Let X be a Banach space. Then $\overline{\mathfrak{F}}$ and $\mathfrak{LC}(X)$ are closed two-sided ideals of $\mathfrak{L}(X)$, so we can consider $\mathfrak{L}(X)/\overline{\mathfrak{F}}$ and $\mathfrak{L}(X)/\mathfrak{LC}(X)$. The latter is called the *Calkin algebra* of X. If X is a Hilbert space they coincide, and we then have a Banach algebra with involution because $\mathfrak{LC}(X)$ is stable by involution.

For a given Banach algebra A there is a particular two-sided ideal of A called the radical or the Jacobson radical of A which plays a very important rôle.

By convention a left or right ideal in a ring is different from the ring.

LEMMA 3.1.1 (W. KRULL). *Let A be a ring with unit 1. Every left (resp. right) ideal of A is contained in a maximal left (resp. right) ideal.*

PROOF. Let L_0 be such a left ideal and let $\mathfrak{F}$ be the family of all left ideals containing L_0. This family is partially ordered by inclusion. This order is inductive because if $\mathfrak{C}$ is a chain in $\mathfrak{F}$ then it is easy to see that either $\bigcup_{I \in \mathfrak{C}} I$ is a left ideal or $\bigcup_{I \in \mathfrak{C}} I = A$. But this last case is impossible, because $1 \notin I$ for all $I \in \mathfrak{C}$. By Zorn's lemma there exists a maximal left ideal in $\mathfrak{F}$. $\square$

LEMMA 3.1.2 (N. JACOBSON). *Let A be an algebra with unit 1 and let $x, y \in A$, $\lambda \in \mathbf{C}$, with $\lambda \neq 0$. Then $\lambda 1 - xy$ is invertible in A if and only if $\lambda 1 - yx$ is invertible in A.*

PROOF. Suppose that $\lambda 1 - xy$ has an inverse u in A. Then we have $(\lambda 1 - xy)u = u(\lambda 1 - xy) = 1$. So we have:

$$(\lambda 1 - yx)(yux + 1) = \lambda yux + \lambda 1 - y(xy)ux - yx$$

$$= \lambda yux + \lambda 1 - y(\lambda u - 1)x - yx = \lambda 1,$$

$$(yux + 1)(\lambda 1 - yx) = \lambda yux + \lambda 1 - yu(xy)x - yx$$

$$= \lambda yux + \lambda 1 - y(\lambda u - 1)x - yx = \lambda 1.$$

Consequently $\lambda 1 - yx$ is invertible in A. $\square$

THEOREM 3.1.3.   *Let A be a ring with unit 1. Then the following sets are identical:*

(i) *the intersection of all maximal left ideals of A,*

(ii) *the intersection of all maximal right ideals of A,*

(iii) *the set of x such that $1 - zx$ is invertible in A, for all $z \in A$,*

(iv) *the set of x such that $1 - xz$ is invertible in A, for all $z \in A$.*

PROOF.   By Lemma 3.1.2, (iii) and (iv) are equivalent. We only prove the equivalence of (i) and (iii), as the equivalence of (ii) and (iv) can be proved by a similar argument. Let x be in the intersection of all maximal left ideals of A. If $a = 1 - zx$ is not invertible then Aa is a left ideal of A so, by Lemma 3.1.1, it is contained in some maximal left ideal L_0. Then $zx \in L_0$ and $1 - zx \in L_0$, so $1 \in L_0$, that is $L_0 = A$ which is a contradiction. Conversely, suppose that $1 - zx$ is invertible for all $z \in A$. If x is not in the intersection of all maximal left ideals, it means that there exists a maximal left ideal L_0 such that $x \notin L_0$. Then $L_0 + Ax = A$ and consequently $1 - zx \in L_0$, but this gives a contradiction because it implies that we have $1 \in L_0$. □

This two-sided ideal of A having properties (i)-(iv) is called the *radical of A* and denoted by Rad A. If Rad $A = \{0\}$ we say that A is *semi-simple*. Of course simple rings (like $M_n(\mathbf{C})$) are semi-simple. There are many examples of semi-simple Banach algebras, for instance examples 1, 2, 3.

THEOREM 3.1.4.   *Let X be a Banach space. Then $\mathcal{L}(X)$ is semi-simple.*

PROOF.   Let $\xi \neq 0$ be fixed in X and let $I_\xi = \{T : T \in \mathcal{L}(X), T\xi = 0\}$. It is obviously a left ideal of $\mathcal{L}(X)$. We prove that it is a maximal left ideal. Suppose that $\mathfrak{J}$ is another left ideal containing I_ξ, with $I_\xi \neq \mathfrak{J}$. Then $\mathfrak{J}\xi = \{T\xi : T \in \mathfrak{J}\}$ is a linear subspace of X different from $\{0\}$ which is invariant under all $S \in \mathcal{L}(X)$. If $\mathfrak{J}\xi \neq X$, let $\eta_1 \in \mathfrak{J}\xi$, $\eta_1 \neq 0$, and $\eta_2 \notin \mathfrak{J}\xi$. Then there exists $S \in \mathcal{L}(X)$ such that $S\eta_1 = \eta_2$. But this contradicts the invariance of $\mathfrak{J}\xi$. So $\mathfrak{J}\xi = X$, and consequently there exists $U \in \mathfrak{J}$ such that $U\xi = \xi$. For an arbitrary $T \in \mathcal{L}(X)$ we have $TU - T \in I_\xi$, so $T \in \mathfrak{J} + I_\xi \subset \mathfrak{J}$, and hence $\mathcal{L}(X) = \mathfrak{J}$. Consequently Rad $\mathcal{L}(X) \subset \bigcap_{\xi \neq 0} I_\xi = \{0\}$. □

Later on we shall give other characterizations of the radical.

Even if a ring is not semi-simple it is always possible to suppose we are in this situation using the following theorem:

THEOREM 3.1.5. *Let A be a ring with unit 1. Then $A/\operatorname{Rad} A$ is semi-simple. Moreover if $x \in A$ the coset $\dot{x}$ is invertible in $A/\operatorname{Rad} A$ if and only if x is invertible in A.*

PROOF. Let L' be a maximal left ideal of $A/\operatorname{Rad} A$. Then $L = \{x : x \in A, \dot{x} \in L'\}$ is a left ideal of A. It is maximal because if J is a left ideal containing L, then $L' \subset \dot{J} = \{\dot{x} : x \in J\}$, so that $L' = \dot{J}$ and thus $L = J$. Conversely, if L is a maximal left ideal of A, then $\dot{L}$ is a maximal left ideal of $A/\operatorname{Rad} A$ for similar reasons. By property (i) this implies that $\{x : \dot{x} \in \operatorname{Rad}(A/\operatorname{Rad} A)\}$ is in the intersection of all maximal left ideals of A, that is the radical of A. Hence $A/\operatorname{Rad} A$ is semi-simple. It is obvious that if x is invertible in A then $\dot{x}$ is invertible in $A/\operatorname{Rad} A$. So suppose the latter property is true. There exists $y \in A$ and $u, v \in \operatorname{Rad} A$ such that $xy - 1 = u$, $yx - 1 = v$. By property (iii), $1 + u$ and $1 + v$ are invertible in A, so we have $xy(1 + u)^{-1} = 1$ and $(1 + v)^{-1}yx = 1$. Hence x is invertible. $\square$

§2. Invertible Elements and Spectrum

Let A be a Banach algebra. We denote by $G(A)$ the set of invertible elements of A. It is obviously a group containing the unit. We now prove that it is an open subset of A.

THEOREM 3.2.1. *Suppose that A is a Banach algebra, $x \in A$ and $\|x\| < 1$. Then $1 - x$ is invertible and*

$$(1 - x)^{-1} = \sum_{k=0}^{\infty} x^k.$$

PROOF. Let $\|x\| = r < 1$ and let $s_n = \sum_{k=0}^{n} x^k$. Obviously, for $n < m$, we have $\|s_n - s_m\| \leq \sum_{k=n+1}^{m} \|x\|^k \leq \frac{r^{n+1}}{1-r}$, so (s_n) is a Cauchy sequence converging to some element $a = \sum_{k=0}^{\infty} x^k$. Because $xs_n = s_{n+1} - 1$ we conclude that $a(1 - x) = (1 - x)a = 1$. $\square$

COROLLARY 3.2.2. *Every left (resp. right, resp. two-sided) ideal of A is disjoint from the open unit ball with centre at 1. Consequently every maximal left (resp. right, resp. two-sided) ideal of A is closed. In particular $\operatorname{Rad} A$ is closed, so $A/\operatorname{Rad} A$ is a semi-simple Banach algebra.*

PROOF. Let L be a maximal left ideal of A. Then L contains no invertible elements, so the intersection of L with the open unit ball centred at 1 is empty by Theorem 3.2.1. Consequently $\overline{L} \cap B(1,1) = \emptyset$ and so $\overline{L} \neq A$. This implies that $L = \overline{L}$ if L is maximal. By Theorem 3.1.3 (i), $\operatorname{Rad} A$ is closed. $\square$

THEOREM 3.2.3. *Suppose that A is a Banach algebra and that a is invertible. If $\|x - a\| < 1/\|a^{-1}\|$, then x is invertible. Moreover the mapping $x \mapsto x^{-1}$ is a homeomorphism from $G(A)$ onto $G(A)$.*

PROOF.   We have $x = a + x - a = a(1 + a^{-1}(x - a))$. We have $\|a^{-1}(x - a)\| \leq \|x - a\| \cdot \|a^{-1}\| < 1$ so, by Theorem 3.2.1, $1 + a^{-1}(x - a)$ is invertible, and consequently x is invertible. Moreover $x^{-1} = (1 + a^{-1}(x - a))^{-1}a^{-1} = \sum_{k=0}^{\infty}(a^{-1}(a - x))^k a^{-1}$. Consequently, we have

$$\|x^{-1} - a^{-1}\| \leq \|a^{-1}\|^2\|x - a\|\sum_{k=0}^{\infty}(\|a^{-1}\| \cdot \|x - a\|)^k .$$

So $x \mapsto x^{-1}$ is continuous, and since it is its own inverse, it is a homeomorphism. $\blacksquare$

As with operators we can define the *spectrum* of x, denoted by $\mathrm{Sp}\, x$, as the set of $\lambda \in \mathbf{C}$ such that $\lambda 1 - x$ is not invertible in A. Also we define the *spectral radius* by $\rho(x) = \sup\{|\lambda| : \lambda \in \mathrm{Sp}\, x\}$.

We have $\lambda 1 - x = \lambda(1 - x/\lambda)$, for $\lambda \neq 0$. So Theorem 3.2.1 implies that $\lambda 1 - x$ is invertible for $\|x\| < |\lambda|$, and consequently $\mathrm{Sp}\, x$ is a bounded subset of $\mathbf{C}$. In other words, it says that $\rho(x) \leq \|x\|$.

Lemma 3.1.2 can be reformulated as saying that $\mathrm{Sp}(xy) \cup \{0\} = \mathrm{Sp}(yx) \cup \{0\}$ and consequently $\rho(xy) = \rho(yx)$, for all $x, y \in A$. Also Theorem 3.1.3 is equivalent to saying that the radical of A is the set of $x \in A$ such that $\rho(xz) = 0$ for all $z \in A$. In particular it implies that the radical is included in the set of *quasi-nilpotent* elements, that is the set of elements a such that $\rho(a) = 0$. But in general this inclusion is strict (for instance consider $M_n(\mathbf{C})$, $\mathcal{L}(X)$ etc...) and in some cases these two sets are identical (for instance if A is commutative, see the remark following Theorem 4.1.2). Theorem 3.1.5 implies that $\mathrm{Sp}\, x = \mathrm{Sp}\, \dot{x}$, for the coset $\dot{x}$ of x in $A/\mathrm{Rad}\, A$.

THEOREM 3.2.4.   *Let A be a Banach algebra and let $x, y \in A$. Suppose that $xy = 1$ and $yx \neq 1$. Then $\mathrm{Sp}\, x$ and $\mathrm{Sp}\, y$ contain a neighbourhood of 0.*

PROOF.   By hypothesis x is not invertible. Let $p = yx \neq 1$. Then $p^2 = y(xy)x = p$. Moreover $(x - \lambda 1)y = xy - \lambda y = 1 - \lambda y$ and $y(x - \lambda 1) = p - \lambda y \neq 1 - \lambda y$. If $1 - \lambda y$ is invertible, then

$$(x - \lambda 1)y(1 - \lambda y)^{-1} = 1.$$

Because y and $(1 - \lambda y)^{-1}$ commute we have

$$y(1 - \lambda y)^{-1}(x - \lambda 1) = (1 - \lambda y)^{-1}y(x - \lambda 1) = (1 - \lambda y)^{-1}(p - \lambda y) \neq 1$$

and consequently $x - \lambda 1$ is not invertible, that is $\lambda \in \mathrm{Sp}\, x$. So we have proved that $B(0, 1/\rho(y)) \subset \mathrm{Sp}\, x$. The argument for $\mathrm{Sp}\, y$ is similar. $\square$

In 1981, S. Berberian and I. Halperin, using a complicated method, proved the following corollary which is in fact an immediate consequence of Theorem 2.2.10 and Theorem 3.2.4.

COROLLARY 3.2.5 (S. BERBERIAN-I. HALPERIN). *Let X be a Banach space and let $T, S \in \mathfrak{L}(X)$ be such that $TS = I$ and $ST \neq I$. Then for every invertible operator $U \in \mathfrak{L}(X)$, both operators $T - U$ and $S - U$ are not compact.*

PROOF. Suppose $T - U = K$ is compact for some U invertible. Then $U^{-1}T - I = U^{-1}K = K_1$ is compact. Taking $T_1 = U^{-1}T$ and $S_1 = SU$ we have $T_1 S_1 = U^{-1}(TS)U = I$ and $S_1 T_1 = ST \neq I$. So by Theorem 3.2.4, $\mathrm{Sp}\, T_1$ contains a disk, but this is absurd by Theorem 2.2.10 since $\mathrm{Sp}\, T_1 = \mathrm{Sp}(I + K_1) = 1 + \mathrm{Sp}\, K_1$ is discrete. $\square$

THEOREM 3.2.6. *Let A be a Banach algebra and let $x \in A$. Then for every non-constant polynomial p with complex coefficients we have $\mathrm{Sp}\, p(x) = p(\mathrm{Sp}\, x)$.*

PROOF. Let $q(z) = C(z - z_1) \ldots (z - z_n)$ be a given arbitrary polynomial. Then $q(x) = C(x - z_1 1) \ldots (x - z_n 1)$. So $q(x)$ is invertible if and only if each of the $x - z_i 1$ is invertible. Applying this to $q(z) = p(z) - \lambda$, we get the result. $\square$

We say that $x \in A$ is *algebraic of degree n* if there exists a polynomial p of degree n such that $p(x) = 0$, and if $q(x) \neq 0$ for all non-zero polynomials q of degree $\leq n - 1$. Such a p with leading coefficient 1 is called a *minimal polynomial* for x (in fact it is unique!).

THEOREM 3.2.7. *Given an integer $n \geq 1$, the set E_n of algebraic elements of A with degree $\leq n$ is closed in A.*

PROOF. Let (x_k) be a sequence of E_n converging to x_0 and for each x_k let p_k be its minimal polynomial. We define a sequence of polynomials q_k of degree n by:

$$q_k(z) = z^{n - \deg p_k} p_k(z).$$

Then the q_k can be written:

$$q_k(z) = \alpha_k^0 + \alpha_k^1 z + \cdots + \alpha_k^{n-1} z^{n-1} + z^n.$$

Let $\beta_k^1, \ldots, \beta_k^n$ be the n roots of q_k. We want to prove that $|\beta_k^i| \le \|x_k\|$. If $\beta_k^i = 0$, this is obvious. Otherwise β_k^i is a root of p_k. If it is not in $\operatorname{Sp} x_k$ then $x_k - \beta_k^i 1$ is invertible, so we have $r(x_k) = 0$, where $r(z) = \frac{p_k(z)}{z - \beta_k^i}$ is a polynomial of degree less than the degree of x_k, which is absurd. So $\beta_k^i \in \operatorname{Sp} x_k$ and then $|\beta_k^i| \le \|x_k\|$. But we have

$$\begin{cases} \alpha_k^{n-1} = -(\beta_k^1 + \ldots + \beta_k^n) \\ \alpha_k^{n-2} = \beta_k^1 \beta_k^2 + \ldots + \beta_k^{n-1}\beta_k^n \\ \quad \vdots \\ \alpha_k^0 = (-1)^n \beta_k^1 \ldots \beta_k^n. \end{cases}$$

So the sequences $(\alpha_k^0), \ldots, (\alpha_k^{n-1})$ are bounded, because (x_k) converges to x_0. Using the Bolzano-Weierstrass theorem, without loss of generality we may suppose that they respectively converge to $\alpha_0, \ldots, \alpha_{n-1}$. Then by continuity we have $\alpha_0 1 + \alpha_1 x_0 + \ldots + \alpha_{n-1} x_0^{n-1} + x_0^n$. So x_0 is algebraic of degree $\le n$. $\blacksquare$

THEOREM 3.2.8 (I.M. GELFAND). *Let A be a Banach algebra and $x \in A$. Then*

(i) $\lambda \to (\lambda 1 - x)^{-1}$ is analytic on $\mathbb{C} \backslash \operatorname{Sp} x$ and goes to 0 at infinity,

(ii) $\operatorname{Sp} x$ is compact and non-empty,

(iii) $\rho(x) = \lim_{n \to \infty} \|x^n\|^{1/n}$.

PROOF. We know that $\rho(x) \le \|x\|$. Moreover $\operatorname{Sp} x$ is closed by Theorem 3.2.3, because $\lambda_0 \notin \operatorname{Sp} x$ and $|\lambda - \lambda_0| < 1/\|(\lambda_0 1 - x)^{-1}\|$ implies $\lambda \notin \operatorname{Sp} x$. So $\operatorname{Sp} x$ is compact. Again by Theorem 3.2.3, $\lambda \to R(\lambda) = (\lambda 1 - x)^{-1}$ is continuous on the open set $\mathbb{C} \backslash \operatorname{Sp} x$. Let $\lambda, \mu \notin \operatorname{Sp} x$. We have $(\lambda 1 - x) - (\mu 1 - x) = (\lambda - \mu)1$, so multiplying by $R(\lambda)R(\mu)$ we get $R(\mu) - R(\lambda) = (\lambda - \mu)R(\lambda)R(\mu)$. Consequently $\lim_{\mu \to \lambda} \frac{R(\mu) - R(\lambda)}{\mu - \lambda} = -R(\lambda)^2$ by continuity of R. So R is analytic on $\mathbb{C} \backslash \operatorname{Sp} x$. Moreover $(\lambda 1 - x)^{-1} = \frac{1}{\lambda}(1 - x/\lambda)^{-1}$ for $\lambda \notin \operatorname{Sp} x$. So, for $|\lambda| \ge 2\|x\|$, we have by Theorem 3.2.1

$$\|(\lambda 1 - x)^{-1}\| \le \frac{2}{|\lambda|}.$$

Hence it goes to zero at infinity. We now prove that $\operatorname{Sp} x$ is non-empty. Suppose this is false and let f be a bounded linear functional on A. Then by (i), $\lambda \mapsto f((\lambda 1 - x)^{-1})$ is entire and goes to zero at infinity so, by Liouville's theorem, it is identically zero. This being true for all $f \in A'$, by Corollary 1.1.4, we have $(\lambda 1 - x)^{-1} = 0$ for all $\lambda \notin \operatorname{Sp} x$, but this is absurd.

We now prove (iii). We know that $\rho(x) \leq \|x\|$, for all $x \in A$. So, by Lemma 3.2.6 applied to z^n, we conclude that

$$\rho(x)^n \leq \|x^n\|. \tag{1}$$

Let $f \in A'$. Then $\lambda \mapsto f((\lambda 1 - x)^{-1})$ is holomorphic on $\mathbb{C} \backslash \operatorname{Sp} x$ by part (i), so in particular if $|\lambda| > \rho(x)$. For $|\lambda| > \|x\|$, by Theorem 3.2.1, we have

$$f((\lambda 1 - x)^{-1}) = \frac{1}{\lambda}\left(f(1) + \frac{f(x)}{\lambda} + \ldots + \frac{f(x^n)}{\lambda^n} + \ldots\right). \tag{2}$$

Consequently this is also true for $|\lambda| > \rho(x)$ by the Identity Principle. Let λ be fixed such that $|\lambda| > \rho(x)$. Then for every $f \in A'$ we have $\sup_n \left|\frac{f(x^n)}{\lambda^n}\right| < +\infty$. By Theorem 1.1.13 applied to A' and to the sequence of $T_n : A' \to \mathbb{C}$, defined by $T_n(f) = \frac{f(x^n)}{\lambda^n}$, we conclude that there exists a constant C, depending on λ, such that $\|x^n\| \leq C|\lambda|^n$ for all $n \geq 1$. Then

$$\limsup_{n \to \infty} \|x^n\|^{1/n} \leq |\lambda| \quad \text{for all } |\lambda| \geq \rho(x). \tag{3}$$

So finally, using (1) and (3), we get:

$$\rho(x) \leq \liminf_{n \to \infty} \|x^n\|^{1/n} \leq \limsup_{n \to \infty} \|x^n\|^{1/n} \leq \rho(x),$$

and the theorem is proved. $\square$

The formula given by (iii) is called the *Beurling-Gelfand formula* mainly because it was used intensively by I.M. Gelfand in 1939 and introduced a bit earlier by A. Beurling in harmonic analysis. This formula is very useful. We gave three applications in Chapter 2 (the examples following Theorem 2.2.10, Theorem 2.2.11 and Corollary 2.3.2) but throughout this book we shall encounter a great number of other applications. This formula can be proved using various other methods.

COROLLARY 3.2.9 (I.M.GELFAND-S.MAZUR). *If A is a Banach algebra in which every non-zero element is invertible then A is isometrically isomorphic to $\mathbb{C}$.*

PROOF. Let $x \in A$. By Theorem 3.2.8 (ii), $\operatorname{Sp} x$ is non-empty. Let $\lambda \in \operatorname{Sp} x$. Then $x - \lambda 1$ is not invertible, consequently $x = \lambda 1$. This implies that $\operatorname{Sp} x$ contains only one point, which we call $\alpha(x)$. The formula $x = \alpha(x)1$ implies that α is an isomorphism from A onto $\mathbb{C}$. It is an isometry because we have $\|x\| = |\alpha(x)| \cdot \|1\| = |\alpha(x)|$. $\square$

REMARK. If $x \in A$ satisfies $q(x) = 0$ for some polynomial q then by Theorem 3.2.6 the spectrum of x is included in the set of zeros of q. In particular, if p is a projection we have $\operatorname{Sp} p \subset \{0, 1\}$. But if p is a non-trivial projection, that is $p \neq 0, 1$, we cannot have $\operatorname{Sp} p = \{0\}$ because $\rho(p) = \lim \|p^n\|^{1/n} = \lim \|p\|^{1/n} = 1$, nor can we have $\operatorname{Sp} p = \{1\}$, because $\rho(1 - p) = 1$ and $\operatorname{Sp}(1 - p) = \{0\}$. So we have $\operatorname{Sp} p = \{0, 1\}$ for a non-trivial projection p.

COROLLARY 3.2.10. *Let A be a Banach algebra. Suppose that $x, y \in A$ satisfy $xy = yx$. Then $\rho(x + y) \leq \rho(x) + \rho(y)$ and $\rho(xy) \leq \rho(x)\rho(y)$.*

PROOF. Because $(xy)^n = x^n y^n$ for every integer $n \geq 1$, using Theorem 3.2.8 (iii), we conclude that

$$\rho(xy) = \lim_{n \to \infty} \|(xy)^n\|^{1/n} \leq \lim_{n \to \infty} \|x^n\|^{1/n} \lim_{n \to \infty} \|y^n\|^{1/n} = \rho(x)\rho(y).$$

Let $\alpha > \rho(x)$, $\beta > \rho(y)$ and $a = x/\alpha$, $b = y/\beta$. Then $\rho(a) < 1$ and $\rho(b) < 1$. So there exists some integer N such that $n \geq N$ implies $\max(\|a^{2^n}\|, \|b^{2^n}\|) < 1$. Defining

$$\gamma_n = \max_{0 \leq k \leq 2^n} \|a^k\| \cdot \|b^{2^n - k}\|,$$

we have

$$\|(x + y)^{2^n}\|^{1/2^n} = \left\| \sum_{k=0}^{2^n} \binom{2^n}{k} x^k y^{2^n - k} \right\|^{1/2^n}$$

$$\leq \left[\sum_{k=0}^{2^n} \binom{2^n}{k} \alpha^k \beta^{2^n - k} \|a^k\| \cdot \|b^{2^n - k}\| \right]^{1/2^n}$$

$$\leq (\alpha + \beta)\gamma_n^{1/2^n}.$$

But it is easy to see that the sequence (γ_n) is decreasing for $n \geq N$, because

$$\max_{0 \leq k \leq 2^{n+1}} \|a^k\| \cdot \|b^{2^{n+1} - k}\|$$

$$= \max \left(\max_{0 \leq k \leq 2^n} \|a^k\| \cdot \|b^{2^{n+1} - k}\|, \max_{2^n \leq k \leq 2^{n+1}} \|a^k\| \cdot \|b^{2^{n+1} - k}\| \right)$$

$$\leq \gamma_n \max(\|a^{2^n}\|, \|b^{2^n}\|).$$

So we have $\rho(x + y) = \lim_{n \to \infty} \|(x + y)^{2^n}\|^{1/2^n} \leq (\alpha + \beta) \limsup_{n \to \infty} \gamma_n^{1/2^n} \leq (\alpha + \beta) \limsup_{n \to \infty} \gamma_N^{1/2^n} = \alpha + \beta$, for arbitrary $\alpha > \rho(x)$, $\beta > \rho(y)$. Hence the theorem is proved. $\square$

Using Gelfand's theory (Chapter IV) it is possible to give an easier proof (see Exercise IV.3).

Suppose that A is a Banach algebra and that B is a closed subalgebra containing the same unit 1. If $x \in B$, what is the relation between the spectrum of x related to B, denoted by $\mathrm{Sp}_B x$, and the spectrum of x related to A, denoted by $\mathrm{Sp}_A x$? Of course, we have $\mathrm{Sp}_A x \subset \mathrm{Sp}_B x$, but is it possible to say more?

THEOREM 3.2.11. *Let (x_n) be a sequence of invertible elements of A converging to a non-invertible element. Then $\lim_{n \to \infty} \|x_n^{-1}\| = +\infty$.*

PROOF. Suppose the result is false. Then there exist $C > 0$ and a subsequence of (x_n), which we denote in the same way, such that $\|x_n^{-1}\| \le C$. Let x be the limit of the sequence (x_n). We have

$$x = x_n(1 + x_n^{-1}(x - x_n))$$

and $\|x_n^{-1}(x - x_n)\| \le C\|x - x_n\| < 1$ for n large enough so, by Theorem 3.2.1, $1 + x_n^{-1}(x - x_n)$ is invertible, and hence x is invertible. So we get a contradiction. $\square$

COROLLARY 3.2.12. *Let $x \in A$ and let α be in the boundary of $\mathrm{Sp}\, x$. Then there exists a sequence (x_n) of elements of A such that $\|x_n\| = 1$ and $\lim_{n \to \infty}(x - \alpha 1)x_n = \lim_{n \to \infty} x_n(x - \alpha 1) = 0$.*

PROOF. Because $\alpha \in \partial \mathrm{Sp}\, x$, there exists a sequence (α_n) of complex numbers converging to α with $\alpha_n \notin \mathrm{Sp}\, x$. We take $x_n = \frac{(x - \alpha_n 1)^{-1}}{\|(x - \alpha_n 1)^{-1}\|}$. Then we have $(x - \alpha 1)x_n = (x - \alpha_n 1)x_n - (\alpha - \alpha_n)x_n = \frac{1}{\|(x - \alpha_n 1)^{-1}\|} - (\alpha - \alpha_n)x_n$. Consequently $\|(x - \alpha 1)x_n\| \le \frac{1}{\|(x - \alpha_n 1)^{-1}\|} + |\alpha - \alpha_n|$. By Theorem 3.2.11, $\lim_{n \to \infty} \|(x - \alpha_n 1)^{-1}\| = +\infty$, so $\lim_{n \to \infty} \|(x - \alpha 1)x_n\| = 0$. The other result is proved similarly. $\square$

In this situation we say that $x - \alpha 1$ is a *topological divisor of zero*. So all boundary points of the spectrum correspond to topological divisors of zero (but the converse is not true in general!).

THEOREM 3.2.13. *Let A be a Banach algebra and let B be a closed subalgebra containing the unit 1. We have:*

(i) *$G(B)$ is the union of some components of $B \cap G(A)$, and furthermore the set $\partial G(B) \cap B \cap G(A)$ is empty,*

(ii) *if $x \in B$, then $\mathrm{Sp}_B x$ is the union of $\mathrm{Sp}_A x$ and a (possibly empty) collection of bounded components of $\mathbf{C} \backslash \mathrm{Sp}_A x$, in particular $\partial \mathrm{Sp}_B x \subset \partial \mathrm{Sp}_A x$.*

PROOF.   (i) It is obvious that $G(B) \subset G(A)$, so $G(B) \subset B \cap G(A)$. These last two sets are open in B. We prove that $B \cap G(A)$ contains no boundary point x of $G(B)$. If such an $x = \lim x_n$, with $x_n \in G(B)$, exists and is in $G(A)$, by continuity of $x \mapsto x^{-1}$ on $G(A)$ we conclude that $x^{-1} = \lim x_n^{-1}$. In particular $(\|x_n^{-1}\|)$ is bounded so, by Theorem 3.2.11, x is not in the boundary of $G(B)$, a contradiction. Let Ω be a component of $B \cap G(A)$ that intersects $G(B)$, and let U be the complement of $\overline{G(B)}$. Since $B \cap G(A)$ contains no boundary points of $G(B)$, Ω is the union of the two open sets $\Omega \cap G(B)$ and $\Omega \cap U$. But Ω is connected, so $\Omega \cap U$ is empty. Hence $\Omega \subset G(B)$. So (i) is proved.

(ii) Let $\Omega_A = \mathbf{C} \backslash \mathrm{Sp}_A x$ and $\Omega_B = \mathbf{C} \backslash \mathrm{Sp}_B x$. Obviously $\Omega_B \subset \Omega_A$. A similar argument shows that $\partial \Omega_B \cap \Omega_A = \emptyset$, so that Ω_B is the union of some components of Ω_A. Hence (ii) is proved. $\square$

COROLLARY 3.2.14.   *With the same hypotheses suppose that $\mathrm{Sp}_A x$ does not separate the complex plane. Then $\mathrm{Sp}_B x = \mathrm{Sp}_A x$.*

This corollary will be used later when $\mathrm{Sp}_A x \subset \mathbf{R}$ or when $\mathrm{Sp}_A x$ is finite or countable.

§3. Holomorphic Functional Calculus

If x is in a Banach algebra A and if $p(\lambda) = \alpha_0 + \alpha_1 \lambda + \cdots + \alpha_n \lambda^n$ is a polynomial with complex coefficients, we can define without any problem $p(x) = \alpha_0 1 + \alpha_1 x + \ldots + \alpha_n x^n$. Now if $r(\lambda) = \frac{p(\lambda)}{q(\lambda)}$ (where p and q are relatively prime) is a rational function with poles outside $\mathrm{Sp}\, x$ then, by Theorem 3.2.6, $q(x)$ is invertible, so we can define $r(x)$ by $p(x)q(x)^{-1}$. Is it possible to extend such definition to a larger class of functions?

Suppose that f is holomorphic on the disk $|\lambda| < R$ and $\rho(x) < R$, so that $f(\lambda) = \sum_{k=0}^{\infty} \alpha_k \lambda^k$ for $|\lambda| < R$. This implies in particular that $\sum_{k=0}^{\infty} \alpha_k \|x^k\|$ converges, so that $f(x) = \sum_{k=0}^{\infty} \alpha_k x^k$ converges in A to an element we call $f(x)$. This argument can be applied in particular to define e^x, the exponential of x.

But if f is only holomorphic on a neighbourhood of $\mathrm{Sp}\, x$, is it possible to define $f(x)$? Of course this function cannot be defined by series because there is a problem in glueing the local pieces. But we now show that it can be done using the Cauchy formula for contours. This idea originates from the pioneering work of F. Riesz.

Let K be a compact subset of $\mathbf{C}$ supporting a measure μ and let f be a continuous function from K into a Banach algebra A. Then, exactly as is done in the scalar situation, it is possible to define $\int_K f(\lambda) \, d\mu(\lambda)$, and of course we have $\phi\left(\int_K f(\lambda) \, d\mu(\lambda)\right) = \int_K \phi(f(\lambda)) \, d\mu(\lambda)$ for all bounded linear functional ϕ on A (see for instance [7], pp. 73–78).

Let $x \in A$ be fixed and let Ω be an arbitrary open set containing $\mathrm{Sp}\, x$. Let Γ be a smooth oriented contour contained in $\Omega \setminus \mathrm{Sp}\, x$ and surrounding $\mathrm{Sp}\, x$. The set $\mathrm{Sp}\, x$ may be very complicated, but we may suppose that Ω is a finite union of polygonal closed curves, without multiple points, and such that $\mathrm{Sp}\, x$ is contained in the union of the bounded components they limit.

For $f \in H(\Omega)$, the algebra of holomorphic functions on Ω, the integral $\frac{1}{2\pi i} \int_\Gamma f(\lambda)(\lambda 1 - x)^{-1} \, d\lambda$ is well-defined because $\lambda \mapsto (\lambda 1 - x)^{-1}$ is defined and continuous on Γ. Moreover this integral is independent of the contour Γ surrounding $\mathrm{Sp}\, x$ for the following reasons.

Suppose that there exist two contours Γ_1, Γ_2 in $\Omega \setminus \mathrm{Sp}\, x$, surrounding $\mathrm{Sp}\, x$ such that

$$x_1 = \frac{1}{2\pi i} \int_{\Gamma_1} f(\lambda)(\lambda 1 - x)^{-1} \, d\lambda \neq \frac{1}{2\pi i} \int_{\Gamma_2} f(\lambda)(\lambda 1 - x)^{-1} \, d\lambda = x_2 \, .$$

By Theorem 1.1.3 there exists $\phi \in A'$ such that $\phi(x_1) \neq \phi(x_2)$. We have

$$\phi(x_1) = \frac{1}{2\pi i} \int_{\Gamma_1} h(\lambda) \, d\lambda \, , \qquad \phi(x_1) = \frac{1}{2\pi i} \int_{\Gamma_2} h(\lambda) \, d\lambda$$

where $h(\lambda) = f(\lambda)\phi((\lambda 1 - x)^{-1})$. But by Theorem 3.2.8 (i), h is holomorphic on $\Omega \setminus \mathrm{Sp}\, x$, so by Cauchy's theorem $\phi(x_1) = \phi(x_2)$ and we get a contradiction. So we can set $f(x) = \frac{1}{2\pi i} \int_\Gamma f(\lambda)(\lambda 1 - x)^{-1} \, d\lambda$ for an arbitrary contour Γ having the previous properties. But before doing that we must at least verify that this definition coincides with the standard one given for rational functions.

LEMMA 3.3.1. *Let $x \in A$, let Γ be a smooth contour surrounding $\mathrm{Sp}\, x$ and let $r(\lambda)$ be a rational function having no poles surrounded by Γ. Then $r(x) = \frac{1}{2\pi i} \int_\Gamma r(\lambda)(\lambda 1 - x)^{-1} \, d\lambda$.*

PROOF. We first prove this for $r(\lambda) = (\alpha - \lambda)^n$, with $n \in \mathbf{Z}$, and α not surrounded by Γ. Let

$$r_n = \frac{1}{2\pi i} \int_\Gamma r(\lambda)(\lambda 1 - x)^{-1} \, d\lambda \, .$$

When $\lambda \notin \mathrm{Sp}\, x$ we have the relation

$$(\lambda 1 - x)^{-1} = (\alpha 1 - x)^{-1} + (\alpha - \lambda)(\alpha 1 - x)^{-1}(\lambda 1 - x)^{-1}.$$

So we have

$$r_n = \frac{1}{2\pi i}(\alpha 1 - x)^{-1} \int_\Gamma (\alpha - \lambda)^n \, d\lambda + (\alpha 1 - x)^{-1} r_{n+1}.$$

But the first integral is zero because $\lambda \mapsto (\alpha - \lambda)^n$ is holomorphic, so we have

$$r_{n+1} = (\alpha 1 - x) r_n. \qquad (1)$$

Consequently the formula will be proved for $(\alpha - \lambda)^n$, $n \in \mathbf{Z}$, if we succeed in proving it for $n = 0$. We have

$$\frac{1}{2\pi i} \int_\Gamma (\lambda 1 - x)^{-1} \, d\lambda = \frac{1}{2\pi i} \int_{\Gamma_R} (\lambda 1 - x)^{-1} \, d\lambda = \frac{1}{2\pi i} \int_{|\lambda|=R} \left(\frac{1}{\lambda} + \frac{x}{\lambda^2} + \cdots \right) d\lambda$$

for $R > \|x\|$. So $\frac{1}{2\pi i} \int_\Gamma (\lambda 1 - x)^{-1} \, d\lambda = 1$. Thus the first part is proved. Now if r is an arbitrary rational function it can be written as

$$r(\lambda) = p(\lambda) + \frac{a_{1,1}}{\alpha_1 - \lambda} + \cdots + \frac{a_{1,k_1}}{(\alpha_1 - \lambda)^{k_1}} + \cdots + \frac{a_{n,1}}{\alpha_n - \lambda} + \cdots + \frac{a_{n,k_n}}{(\alpha_n - \lambda)^{k_n}} \qquad (2)$$

where p is a polynomial in λ and where the $\alpha_1, \ldots, \alpha_n$ are the poles with their respective multiplicities $k_1, \ldots, k_n$ (decomposition in simple elements). A formal calculation implies that we also have

$$r(x) = p(x) + a_{1,1}(\alpha_1 1 - x)^{-1} + \cdots + a_{1,k_1}(\alpha_1 1 - x)^{-k_1} + \cdots$$
$$\cdots + a_{n,1}(\alpha_n 1 - x)^{-1} + \cdots + a_{n,k_n}(\alpha_n 1 - x)^{-k_n}. \qquad (3)$$

So the first part implies the result. $\square$

COROLLARY 3.3.2 (CAYLEY-HAMILTON THEOREM). *Let a be a $n \times n$ complex matrix and let $p(\lambda) = \det(a - \lambda 1)$ be its characteristic polynomial. Then $p(a) = 0$.*

PROOF. Let $\alpha_1, \ldots, \alpha_k$ be the different eigenvalues of a and let Γ be the union of k small disjoint circles with centres respectively at $\alpha_1, \ldots, \alpha_k$. By Lemma 3.3.1 we have

$$p(a) = \frac{1}{2\pi i} \int_\Gamma p(\lambda)(\lambda 1 - a)^{-1} \, d\lambda.$$

But $(a - \lambda 1)^{-1} = \frac{1}{\det(a - \lambda 1)} b(\lambda)$, where $b(\lambda)$ is a $n \times n$ matrix depending analytically on λ since its (i,j) entry is the (j,i) cofactor of $a - \lambda 1$, and so is a polynomial in λ of degree $\leq n - 1$. We then have

$$p(a) = -\frac{1}{2\pi i} \int_\Gamma b(\lambda) d\lambda = 0$$

by Cauchy's theorem. $\square$

THEOREM 3.3.3 (HOLOMORPHIC FUNCTIONAL CALCULUS). *Let A be a Banach algebra and let $x \in A$. Suppose that Ω is an open set containing $\operatorname{Sp} x$ and that Γ is an arbitrary smooth contour included in Ω and surrounding $\operatorname{Sp} x$. Then the mapping $f \mapsto f(x) = \frac{1}{2\pi i} \int_\Gamma f(\lambda)(\lambda 1 - x)^{-1}\, d\lambda$ from $H(\Omega)$ into A has the following properties:*

(i) $(f_1 + f_2)(x) = f_1(x) + f_2(x)$,

(ii) $(f_1 \cdot f_2)(x) = f_1(x) \cdot f_2(x) = f_2(x) \cdot f_1(x)$,

(iii) $1(x) = 1$ and $I(x) = x$ (where $I(\lambda) = \lambda$),

(iv) if (f_n) converges to f uniformly on compact subsets of Ω, then $f(x) = \lim f_n(x)$,

(v) $\operatorname{Sp} f(x) = f(\operatorname{Sp} x)$.

PROOF. (i) is obvious by definition. (iii) follows immediately from Lemma 3.3.1. (iv) is also easy because (f_n) converges uniformly to f on Γ and $\|(\lambda 1 - x)^{-1}\|$ is bounded on this set. We now prove (ii). By Runge's theorem (see [7], p. 288) there exist two sequences of rational functions (r_k^1) and (r_k^2), with poles not surrounded by Γ, converging uniformly respectively to f_1 and f_2. By Lemma 3.3.1 we have $(r_k^1 r_k^2)(x) = r_k^1(x) \cdot r_k^2(x)$. So by (iv), property (ii) is true. We finish by proving (v). If f has no zero on $\operatorname{Sp} x$ then $g = 1/f$ is holomorphic on a neighbourhood Ω_1 of $\operatorname{Sp} x$. If necessary we can replace Γ by a contour $\Gamma_1 \subset \Omega_1$ surrounding $\operatorname{Sp} x$. We have $f(\lambda)g(\lambda) = 1$ on Γ_1 and consequently, applying (ii) and (iii) for Γ_1 and Ω_1, we get $f(x)g(x) = 1$. Thus $f(x)$ is invertible in A. On the other hand, if $f(\alpha) = 0$ for some $\alpha \in \operatorname{Sp} x$, then there exists $h \in H(\Omega)$ such that $f(\lambda) = (\alpha - \lambda)h(\lambda)$ on Ω, and consequently $f(x) = (\alpha 1 - x)h(x)$. But $\alpha 1 - x$ is not invertible, so $f(x)$ is not invertible. Therefore $\beta 1 - f(x)$ is not invertible if and only if f takes the value β on $\operatorname{Sp} x$. $\square$

We now give several elementary applications of this extremely important theorem.

THEOREM 3.3.4. *Let A be a Banach algebra. Suppose that $x \in A$ has a disconnected spectrum. Let U_0, U_1 be two disjoint open sets such that $\operatorname{Sp} x \subset U_0 \cup U_1$, $\operatorname{Sp} x \cap U_0 \neq \emptyset$ and $\operatorname{Sp} x \cap U_1 \neq \emptyset$. Then there exists a non-trivial projection p commuting with x, such that*

$$\operatorname{Sp}(px) = (\operatorname{Sp} x \cap U_1) \cup \{0\}, \quad \operatorname{Sp}(x - px) = (\operatorname{Sp} x \cap U_0) \cup \{0\}.$$

PROOF. We consider the holomorphic function f defined on $U = U_0 \cup U_1$ by $f(\lambda) = 0$ on U_0 and $f(\lambda) = 1$ on U_1. We set $p = f(x)$. Because $f(\lambda)^2 = f(\lambda)$, we get $p^2 = p$. By construction p commutes with x, by Theorem 3.3.3 (ii) and (iii). By part (v) of Theorem 3.3.3 we have $\operatorname{Sp} p = \{0,1\}$, so p is non-trivial. Moreover, we have $px = f(x)I(x) = (fI)(x)$. But $(fI)(\lambda) = \lambda$ on U_1 and $(fI)(\lambda) = 0$ on U_0, so by Theorem 3.3.3 (v) we have $\operatorname{Sp}(px) = (\operatorname{Sp} x \cap U_1) \cup \{0\}$. For $\operatorname{Sp}(x - px)$ the result is obtained similarly. $\square$

THEOREM 3.3.5. *Let A be a Banach algebra. Suppose that $x \in A$ and that $\alpha \notin \operatorname{Sp} x$. Then we have*

$$\operatorname{dist}(\alpha, \operatorname{Sp} x) = \frac{1}{\rho((\alpha \mathbf{1} - x)^{-1})}.$$

PROOF. Let Ω be an open set containing $\operatorname{Sp} x$, but not α. Then $f(\lambda) = 1/(\alpha - \lambda)$ is holomorphic on Ω. So by Theorem 3.3.3 (v) we have

$$\operatorname{Sp}(\alpha \mathbf{1} - x)^{-1} = \left\{ \frac{1}{\alpha - \lambda} : \lambda \in \operatorname{Sp} x \right\}.$$

So in particular,

$$\rho((\alpha \mathbf{1} - x)^{-1}) = \sup \left\{ \frac{1}{|\alpha - \lambda|} : \lambda \in \operatorname{Sp} x \right\}$$
$$= 1/\inf\{|\alpha - \lambda| : \lambda \in \operatorname{Sp} x\} = 1/\operatorname{dist}(\alpha, \operatorname{Sp} x). \ \square$$

THEOREM 3.3.6. *Let A be a Banach algebra. Suppose that $x \in A$ has a spectrum which does not separate 0 from infinity. Then there exists $y \in A$ such that $x = e^y$. In particular, for every integer $n \geq 1$ there exists $z \in A$ such that $z^n = x$.*

PROOF. By hypothesis 0 is in the unbounded component of $\mathbf{C} \setminus \operatorname{Sp} x$. Hence there exists a simply connected open set Ω containing $\operatorname{Sp} x$ and $f \in H(\Omega)$ such that $e^{f(\lambda)} = \lambda$ on Ω. Then we apply Theorem 3.3.3 and take $y = f(x)$. For each $n \geq 1$ we take $z = e^{y/n}$. $\square$

This implies in particular the non-trivial fact that an invertible $n \times n$ matrix is an exponential (obviously the converse is true).

Let A be a Banach algebra. We denote by $\exp(A)$ the set of all products of exponentials $e^{x_1} \cdots e^{x_n}$, where $x_1, \ldots, x_n \in A$. It is obvious that $\exp(A) \subset G(A)$. But $t \mapsto e^{tx_1} \cdots e^{tx_n}$ is a continuous function from $[0,1]$ into $G(A)$ which connects 1 and $e^{x_1} \ldots e^{x_n}$. So in fact $\exp(A)$ is included in the connected component of $G(A)$ containing 1, which is denoted by $G_1(A)$ and is called the *principal component* of $G(A)$.

THEOREM 3.3.7 (E.R. LORCH). *If A is a Banach algebra we have $\exp(A) = G_1(A)$.*

PROOF. First we prove that $\exp(A)$ is open in $G(A)$. Let $a \in \exp(A)$ and suppose that $\|x - a\| < 1/\|a^{-1}\|$. Then $\|1 - a^{-1}x\| = \|a^{-1}(a-x)\| < 1$ and so $\rho(1 - a^{-1}x) < 1$. Consequently $\mathrm{Sp}(a^{-1}x)$ is included in the open disk of centre 1 and radius 1. But for $\mathrm{Re}\,\lambda > 0$ there exists a holomorphic function f such that $e^{f(\lambda)} = \lambda$. So by Theorem 3.3.3 there exists $y = f(a^{-1}x)$ such that $e^y = a^{-1}x$. Then $x = ae^y \in exp(A)$. We now prove that $\exp(A)$ is closed in $G(A)$. If $a_n \in \exp(A)$ and $\lim_{n \to \infty} a_n = a$ then $(a_n^{-1}a)$ converges to 1 and so $\|1 - a_n^{-1}a\| < 1$ for n large. As before, we conclude that $a_n^{-1}a = e^{z_n}$ for some $z_n \in A$, so $a \in \exp(A)$. Because $\exp(A)$ is closed and open in $G(A)$ and is contained in $G_1(A)$, we conclude that $\exp(A) = G_1(A)$. $\square$

For $x \in A$ we define the *exponential spectrum* of x, denoted by $\epsilon(x)$, by the set of $\lambda \in \mathbf{C}$ such that $\lambda 1 - x \notin \exp(A)$. Obviously we have $\mathrm{Sp}\,x \subset \epsilon(x)$. Let $(\mathrm{Sp}\,x)^{\hat{}}$ be the polynomially convex hull of $\mathrm{Sp}\,x$, that is the union of $\mathrm{Sp}\,x$ with the bounded components of $\mathbf{C} \setminus \mathrm{Sp}\,x$, and let $\lambda_0 \notin (\mathrm{Sp}\,x)^{\hat{}}$. Then by Theorem 3.3.6 there exists y such that $\lambda_0 1 - x = e^y$, and consequently $\lambda_0 \notin \epsilon(x)$. In other words, $\mathrm{Sp}\,x \subset \epsilon(x) \subset (\mathrm{Sp}\,x)^{\hat{}}$. This implies in particular that $\epsilon(x)$ is a non-empty compact subset of $\mathbf{C}$ because $G_1(A)$ is open in A. We now prove a simple and nice result characterizing the spectrum of $\dot{x}$, for $\dot{x} \in A/I$, where I is a closed two-sided ideal of A.

THEOREM 3.3.8 (R. HARTE). *Let T be a continuous morphism from a Banach algebra A onto a Banach algebra B. Then $T(\exp(A)) = \exp(B)$. So we have*

$$\epsilon(Tx) = \bigcap_{y \in \mathrm{Ker}\,T} \epsilon(x+y) \quad \text{and} \quad \mathrm{Sp}\,Tx \subset \bigcap_{y \in \mathrm{Ker}\,T} \mathrm{Sp}(x+y) \subset (\mathrm{Sp}\,Tx)^{\hat{}}.$$

In particular, if I is a closed two-sided ideal of A then

$$\mathrm{Sp}\,\dot{x} \subset \bigcap_{y \in I} \mathrm{Sp}(x+y) \subset (\mathrm{Sp}\,\dot{x})^{\hat{}}.$$

PROOF. It is clear that $T(e^{x_1} \cdots e^{x_n}) = e^{Tx_1} \cdots e^{Tx_n}$, so the first part is obvious. If $\lambda \notin \epsilon(Tx)$, then $Tx - \lambda 1 = Tu$ for some $u \in \exp(A)$. So, taking $y = u - x + \lambda 1$, we have $x + y - \lambda \in \exp(A)$ and $Ty = 0$, and consequently $\bigcap_{y \in \mathrm{Ker}\,T} \epsilon(x+y) \subset \epsilon(Tx)$. The other inclusion is obvious. Moreover $\bigcap_{y \in \mathrm{Ker}\,T} \mathrm{Sp}(x+y) \subset \bigcap_{y \in \mathrm{Ker}\,T} \epsilon(x+y) = \epsilon(Tx) \subset (\mathrm{Sp}\,Tx)^{\hat{}}$ by the previous remark. The last part follows immediately, taking $B = A/I$ and $Tx = \dot{x}$. $\square$

§4. Analytic Properties of the Spectrum

Let A be a Banach algebra. The main question in spectral theory is the following: what can be said about the spectrum function $x \mapsto \operatorname{Sp} x$ when x varies in A? Is that function continuous or analytic in some sense? In order to measure the continuity of the spectrum we introduce a distance on the set of compact subsets of $\mathbf{C}$, called the *Hausdorff distance* and defined by

$$\Delta(K_1, K_2) = \max(\sup_{z \in K_2} \operatorname{dist}(z, K_1), \sup_{z \in K_1} \operatorname{dist}(z, K_2))$$

for K_1, K_2 compact subsets of $\mathbf{C}$. Let $r > 0$ and K be a compact subset of $\mathbf{C}$. If $K + r$ denotes $\{ z : \operatorname{dist}(z, K) \leq r \}$ then obviously $K_1 \subset K_2 + \Delta(K_1, K_2)$ and $K_2 \subset K_1 + \Delta(K_1, K_2)$. We shall say that $x \mapsto \operatorname{Sp} x$ is *continuous at* $a \in A$ if, for every $\epsilon > 0$, there exists $\delta > 0$ such that $\|x - a\| < \delta$ implies $\Delta(\operatorname{Sp} x, \operatorname{Sp} a) < \epsilon$. As usual we shall say that $x \mapsto \operatorname{Sp} x$ is *continuous* on E if it is continuous at every point of E. If for a given $\epsilon > 0$, the number $\delta > 0$ is independent of a on E, we say that $x \mapsto \operatorname{Sp} x$ is *uniformly continuous on* E. If E is a cone, that is $\alpha E \subset E$ for all $\alpha > 0$, it is equivalent to say that there exists $C > 0$ such that $\Delta(\operatorname{Sp} x, \operatorname{Sp} y) \leq C\|x - y\|$ for $x, y \in E$.

Of course there are examples of algebras where the spectrum behaves nicely.

THEOREM 3.4.1. *Let A be a Banach algebra. Suppose that $x, y \in A$ commute. Then $\operatorname{Sp} y \subset \operatorname{Sp} x + \rho(x - y)$ and consequently we have $\Delta(\operatorname{Sp} x, \operatorname{Sp} y) \leq \rho(x - y) \leq \|x - y\|$. Furthermore, if A is commutative then the spectrum function is uniformly continuous on A.*

PROOF. Suppose the inclusion is not true. This means that there exists $\alpha \in \operatorname{Sp} y$ such that $\operatorname{dist}(\alpha, \operatorname{Sp} x) > \rho(x - y)$. Therefore by Theorem 3.3.5, we have $\rho((\alpha 1 - x)^{-1})\rho(x - y) < 1$. So, by Corollary 3.2.10, we have $\rho((\alpha 1 - x)^{-1}(x - y)) < 1$. But $\alpha 1 - y = \alpha 1 - x + x - y$ and $\alpha 1 - x$ is invertible by hypothesis, so $\alpha 1 - y = (\alpha 1 - x)[1 + (\alpha 1 - x)^{-1}(x - y)]$ is also invertible, which is a contradiction. $\square$

If $A = M_n(\mathbf{C})$ then the spectrum function is continuous (this derives from the Implicit Function Theorem and also from Corollary 3.4.5, as we shall see below). But in that case it is not uniformly continuous because if we take

$$a_n = \begin{pmatrix} n^2 & 1 \\ n^2(n - n^2) & n - n^2 \end{pmatrix}, \qquad b_n = \begin{pmatrix} n^2 & 1 \\ n^2(n - n^2) & n - n^2 - 1/n \end{pmatrix},$$

then $\|a_n - b_n\| = 1/n$ and $\Delta(\text{Sp } a_n, \text{Sp } b_n) \geq 1/2$, for n large enough.

If $A = \mathfrak{LC}(X)$ then the spectrum function is also continuous (see Corollary 3.4.5) but not uniformly continuous.

In general the spectrum function behaves very badly. Let us consider $A = \mathfrak{L}(H)$, where H is a Hilbert space. Then the spectrum function is uniformly continuous when restricted to the subspace of self-adjoint operators (see Theorem 6.2.1), but it is not continuous for practically all self-adjoint operators: this difficult result comes from a theorem of B.B. Morrel and J. Morrel.

We only give an easier result due to S. Kakutani showing that the spectrum function can be discontinuous.

EXAMPLE. Let (α_n) be the sequence of positive numbers defined by $\alpha_n = e^{-k}$ if $n = 2^k(2\ell + 1)$. Then in the separable Hilbert space H with orthonormal basis (e_n) we consider the weighted shift T with weight (α_n). For every $k \geq 1$ we define $T_k \in \mathfrak{L}(H)$ by

$$T_k e_n = \begin{cases} 0 & \text{, if } n = 2^k(2\ell + 1), \text{ for some } \ell, \\ \alpha_n e_{n+1} & \text{, otherwise.} \end{cases}$$

It is easy to verify that $T_k^{2^{k+1}} e_n = 0$, for every n, so the operators T_k are nilpotent, and hence $\text{Sp } T_k = \{0\}$. We have

$$(T - T_k)e_n = \begin{cases} e^{-k} e_{n+1} & \text{, if } n = 2^k(2\ell + 1), \text{ for some } \ell, \\ 0 & \text{, otherwise.} \end{cases}$$

So $\|T - T_k\| \leq e^{-k}$, and hence the operators T_k converge to T.

We have $T^m e_n = \alpha_n \alpha_{n+1} \cdots \alpha_{n+m-1} e_{n+m}$ and consequently we have $\|T^m\| = \sup_n (\alpha_n \alpha_{n+1} \cdots \alpha_{n+m-1})$. By the definition of the sequence (α_n) we have

$$\alpha_1 \alpha_2 \cdots \alpha_{2^t-1} = \prod_{j=1}^{t-1} \exp(-j 2^{t-j-1}).$$

So $(\alpha_1 \alpha_2 \cdots \alpha_{2^t-1})^{1/(2^t-1)} > \left(\prod_{j=1}^{t-1} \exp(-j/2^{j+1}) \right)^2$. Let $\sigma = \sum_{j=1}^{\infty} j/2^{j+1}$. Then $0 < e^{2\sigma} \leq \lim_{m \to \infty} \|T^m\|^{1/m} = \rho(T)$. So $\text{Sp } T \neq \{0\}$.

Furthermore, V. Müller has given a more sophisticated example where the spectrum function is discontinuous even on the real line. He proved that there exist $T, S \in \mathfrak{L}(H)$ such that $\rho(T) > 0$ and $\rho(T + \lambda S) = 0$ for all rational numbers λ in $]0, 1[$ (see [1], pp. 36–38). We now give some positive results. The first one has been known for a long time.

THEOREM 3.4.2.   *Let A be a Banach algebra. Then the spectrum function $x \mapsto$
$\mathrm{Sp}\, x$ is upper semicontinuous on A, that is for every open set U containing $\mathrm{Sp}\, x$
there exists $\delta > 0$ such that $\|x - y\| < \delta$ implies $\mathrm{Sp}\, y \subset U$.*

PROOF.   Suppose that there exist sequences (y_n) and (α_n) such that $x = \lim_n y_n$,
$\alpha_n \in \mathrm{Sp}\, y_n \cap (\mathbf{C} \backslash U)$. Then $|\alpha_n| \leq \|y_n\|$, so (α_n) is a bounded sequence. By
the Bolzano-Weierstrass theorem we may suppose without loss of generality that
it converges to α. But $\alpha \notin U$ because $\mathbf{C} \backslash U$ is closed, so $\alpha 1 - x$ is invertible. By
Theorem 3.2.3, $\alpha_n 1 - y_n$ will be invertible for n large, which is a contradiction. $\square$

The following theorem is a particular case of an old result of K. Kuratowski.
It says that even if the spectrum function is discontinuous, the set of its points of
continuity is rather large.

THEOREM 3.4.3.   *Let A be a Banach algebra. Then the set of points of continuity
of $x \mapsto \mathrm{Sp}\, x$ is a dense G_δ-subset of A.*

PROOF.   It is easy to see that the algebra of real continuous functions on $\mathbf{C}$ is
separable. So we suppose that (f_n) is a dense sequence of functions in this algebra.
We define $\hat{f}_n(x) = \sup\{f_n(\lambda) : \lambda \in \mathrm{Sp}\, x\}$, for $x \in A$. By Theorem 3.4.2, the $\hat{f}_n$ are
upper semicontinuous in the classical sense.

We now prove that $x \mapsto \mathrm{Sp}\, x$ is continuous at a if and only if the $\hat{f}_n$ are
continuous at a. Let $\lambda_0 \in \mathrm{Sp}\, a$ be such that $f_n(\lambda_0) = \hat{f}_n(a)$ and let $\epsilon > 0$. Because
the function f_n is continuous at λ_0, there exists $\delta > 0$ such that $|\lambda - \lambda_0| < \delta$ implies
$|f_n(\lambda) - f_n(\lambda_0)| < \epsilon$. If the spectrum function is continuous at a, there exists $r > 0$
such that $|x - a| < r$ implies that there exists $\lambda \in \mathrm{Sp}\, x$ such that $|\lambda - \lambda_0| < \delta$. Then
$\hat{f}_n(x) \geq f_n(\lambda) \geq f_n(\lambda_0) - \epsilon$. So, with the upper semicontinuity of $\hat{f}_n$, this implies
the continuity of $\hat{f}_n$. Conversely, suppose that all $\hat{f}_n$ are continuous at a and that
$x \mapsto \mathrm{Sp}\, x$ is discontinuous at a. Then there exist a sequence (x_k) converging to
a, $\alpha \in \mathrm{Sp}\, a$, and $r > 0$ such that $B(\alpha, r) \cap \mathrm{Sp}\, x_k = \emptyset$ for all k. There exists f
continuous on $\mathbf{C}$ such that $f(\alpha) = 1$ and $f(z) = 0$ for $|z - \alpha| \geq r$, so there exists n
such that $f_n(\alpha) \geq 2/3$ and $f_n(z) \leq 1/3$ for $|z - \alpha| \geq r$. Consequently $\hat{f}_n(a) \geq 2/3$
and $\hat{f}_n(x_k) \leq 1/3$ and this is a contradiction because $\hat{f}_n(a) = \lim_k \hat{f}_n(x_k)$.

Let C be the set of points of continuity of $x \mapsto \mathrm{Sp}\, x$ and C_n be the set of points
of continuity of $\hat{f}_n$. So we have $C = \cap_{n=1}^{\infty} C_n$. Using Theorem 1.1.1 it is sufficient
to prove that C_n is a dense G_δ-subset of A. So let f be an upper semicontinuous
function from A into $\mathbf{R}$, let D be its set of points of discontinuity and let $(U_n)_{n \geq 1}$

be a countable basis of **R**. In order to prove the theorem we have only to show that

$$D = \bigcup_{n \geq 1} D_n$$

where D_n is $f^{-1}(U_n)$ minus its interior points. If $x \in D$ there exist some U_n containing $f(x)$ such that $f^{-1}(U_n)$ is not a neighbourhood of x, and so $x \in D_n$. Conversely, if $x \in D_n$, then U_n is a neighbourhood of $f(x)$ such that $f^{-1}(U_n)$ is not a neighbourhood of x, so $x \in D$. Now $D_n \subset \overline{f^{-1}(U_n)} \setminus (f^{-1}(U_n))^0 = \partial f^{-1}(U_n)$, so D_n has no interior points and is a F_σ-set. Hence we get the result. $\square$

The first important results concerning spectral variation are due to J.D. Newburgh. It is amazing to notice that these results are not given in all standard books on Banach algebra theory. They were given publicity for the first time in [1].

THEOREM 3.4.4 (J.D. NEWBURGH). *Let A be a Banach algebra and $x \in A$. Suppose that U, V are two disjoint open sets such that $\operatorname{Sp} x \subset U \cup V$ and $\operatorname{Sp} x \cap U \neq \emptyset$. Then there exists $r > 0$ such that $\|x - y\| < r$ implies $\operatorname{Sp} y \cap U \neq \emptyset$.*

PROOF. By Theorem 3.4.2 there exists $\delta > 0$ such that $\|x - y\| < \delta$ implies $\operatorname{Sp} y \subset U \cup V$. Suppose the theorem to be false. Then there exists a sequence (y_n) converging to x such that $\operatorname{Sp} y_n \subset V$, for n large enough. Let f be the holomorphic function on $U \cup V$ defined by 1 on U and 0 on V. By Holomorphic Functional Calculus we have $\lim f(y_n) = f(x)$ and $f(y_n) = 0$ for n large enough. But $\operatorname{Sp} f(x) = f(\operatorname{Sp} x)$ contains 1, so $f(x) \neq 0$, which gives a contradiction. $\square$

COROLLARY 3.4.5 (J.D. NEWBURGH). *Suppose that the spectrum of a is totally disconnected. Then $x \mapsto \operatorname{Sp} x$ is continuous at a.*

PROOF. Let $\epsilon > 0$. Because $\operatorname{Sp} a$ is totally disconnected it is included in the union U of a finite number of disjoint open sets, intersecting $\operatorname{Sp} a$ and having diameters less than ϵ. By Theorem 3.4.1, there exists $r_1 > 0$ such that $\|x - a\| < r_1$ implies $\operatorname{Sp} x \subset U$. Applying Theorem 3.4.4 to U, there exists $r_2 > 0$ such that $\|x - a\| < r_2$ implies $\sup \operatorname{dist}(z, \operatorname{Sp} x) < \epsilon$ for $z \in \operatorname{Sp} a$. So $\|x - a\| < \min(r_1, r_2)$ implies $\triangle(\operatorname{Sp} a, \operatorname{Sp} x) < \epsilon$. $\square$

This implies in particular that the spectral function is continuous at all elements having finite or countable spectrum.

In 1966, A. Brown and R.G. Douglas considered the problem of formulating a maximum principle for the multifunction $\lambda \mapsto \mathrm{Sp}(f(\lambda))$ where f is an analytic function from a domain D of $\mathbf{C}$ into a Banach algebra A. In 1968–1970, E. Vesentini solved this question by proving Theorems 3.4.7, 3.4.11 and 3.4.13.

Let D be a domain of $\mathbf{C}$. A function ϕ from D into $\mathbf{R} \cup \{-\infty\}$ is said to be *subharmonic* on D if it is upper semicontinuous on D and satisfies the mean inequality

$$\phi(\lambda_0) \le \frac{1}{2\pi} \int_0^{2\pi} \phi(\lambda_0 + re^{i\theta}) \, d\theta$$

for all closed disks $\overline{B}(\lambda_0, r)$ included in D. In particular h is harmonic on D if and only if h and $-h$ are subharmonic on D. Subharmonic functions have a great number of very beautiful properties which we cannot give in detail in this small book. So we refer the reader to the appendix for a quick survey, or to a standard textbook ([4], for instance).

LEMMA 3.4.6. *Let f be an analytic function from a domain D of $\mathbf{C}$ into a Banach space X. Then $\lambda \mapsto \log \|f(\lambda)\|$ is subharmonic on D.*

PROOF. Obviously this function is continuous. Let $\overline{B}(\lambda_0, r)$ be a closed disk included in D. By Cauchy's theorem we have

$$f(\lambda_0) = \frac{1}{2\pi} \int_0^{2\pi} f(\lambda_0 + re^{i\theta}) \, d\theta \,,$$

and consequently

$$\|f(\lambda_0)\| \le \frac{1}{2\pi} \int_0^{2\pi} \|f(\lambda_0 + re^{i\theta})\| \, d\theta \,.$$

For every polynomial p, $|e^{p(\lambda)}| \cdot \|f(\lambda)\| = \|e^{p(\lambda)} f(\lambda)\|$ and $\lambda \mapsto e^{p(\lambda)} f(\lambda)$ is analytic, so, by the first part, $|e^{p(\lambda)}| \cdot \|f(\lambda)\|$ is subharmonic. Then the Beckenbach-Saks theorem (see Theorem A.1.8) implies that $\log \|f(\lambda)\|$ is subharmonic. $\square$

THEOREM 3.4.7 (E. VESENTINI). *Let f be an analytic function from a domain D of $\mathbf{C}$ into a Banach algebra A. Then $\lambda \mapsto \rho(f(\lambda))$ and $\lambda \mapsto \log \rho(f(\lambda))$ are subharmonic on D.*

PROOF. By Theorem 3.2.8 the sequence $\frac{1}{2^n} \log \|f(\lambda)^{2^n}\|$ decreases and converges to $\log \rho(f(\lambda))$. Because $\lambda \mapsto f(\lambda)^{2^n}$ is analytic, then by Lemma 3.4.6 the function $\lambda \mapsto \log \|f(\lambda)^{2^n}\|$ is subharmonic. So $\log \rho(f(\lambda))$, being a decreasing limit of subharmonic functions, is subharmonic. Taking the composition of this function with e^t, which is convex and increasing, we conclude that $\rho(f(\lambda))$ is subharmonic. $\square$

It is fair to say that B. Schmidt and V. Istrăţescu proved Theorem 3.4.7 practically at the same time, but E. Vesentini was the first to obtain non-trivial applications in spectral theory of this nice and simple theorem.

In the rest of this section and mainly in Chapter V, we shall see that Theorem 3.4.7 has a huge number of important applications.

COROLLARY 3.4.8. *Let f be an analytic function from a domain D of $\mathbb{C}$ into a Banach algebra A. Suppose that $\alpha \notin \mathrm{Sp}\, f(\lambda)$ for all $\lambda \in D$. Then $\lambda \mapsto 1/\mathrm{dist}(\alpha, \mathrm{Sp}\, f(\lambda))$ and $\lambda \mapsto -\log \mathrm{dist}(\alpha, \mathrm{Sp}\, f(\lambda))$ are subharmonic on D.*

PROOF. By Theorem 3.3.5 we have $\mathrm{dist}(\alpha, \mathrm{Sp}\, f(\lambda)) = \frac{1}{\rho((\alpha 1 - f(\lambda))^{-1})}$, so we apply Theorem 3.4.7 to the analytic function $\lambda \mapsto (\alpha 1 - f(\lambda))^{-1}$. $\square$

COROLLARY 3.4.9. *Let f be an analytic function from a domain D of $\mathbb{C}$ into a Banach algebra A. Define*

$$u(\lambda) = \max\{\mathrm{Re}\, u : u \in \mathrm{Sp}\, f(\lambda)\},$$
$$v(\lambda) = \min\{\mathrm{Re}\, v : v \in \mathrm{Sp}\, f(\lambda)\}.$$

Then $u(\lambda) = \log \rho(e^{f(\lambda)})$, $v(\lambda) = -\log \rho(e^{-f(\lambda)})$, and u and $-v$ are subharmonic on D.

PROOF. It is obvious that $v(\lambda) = -\max\{\mathrm{Re}\, v : v \in \mathrm{Sp}(-f(\lambda))\}$, so it is sufficient to prove that u is subharmonic. We first prove that $\max\{\mathrm{Re}\, u : u \in \mathrm{Sp}\, x\} = \log \rho(e^x)$, for $x \in A$. By Theorem 3.3.3 applied to $f(z) = e^z$, we have $\log \rho(e^x) = \max\{\log |e^u| : u \in \mathrm{Sp}\, x\} = \max\{\mathrm{Re}\, u : u \in \mathrm{Sp}\, x\}$. Consequently $u(\lambda) = \log \rho(e^{f(\lambda)})$. Then the result comes from Theorem 3.4.7. $\square$

COROLLARY 3.4.10. *Let $0 \le r \le s$, $0 < \theta_2 - \theta_1 < 2\pi$, $\Omega = \{z : |z| > s, \theta_1 < \arg z < \theta_2\}$ and let f be an analytic function from a domain D of $\mathbb{C}$ into a Banach algebra A such that $\mathrm{Sp}\, f(\lambda) \subset \Omega \cup B(0, r)$, for all $\lambda \in D$. Define*

$$u(\lambda) = \max\{\arg z : z \in \mathrm{Sp}\, f(\lambda) \cap \Omega\},$$
$$v(\lambda) = \min\{\arg z : z \in \mathrm{Sp}\, f(\lambda) \cap \Omega\}.$$

Then u and $-v$ are subharmonic on D.

PROOF. Without loss of generality we may suppose that $\mathrm{Sp}\, f(\lambda) \cap \Omega \notin \emptyset$, for all $\lambda \in D$. On Ω we consider the branch of the logarithm $\log z = \log |z| + i \arg z$ and we define

$$h(z) = \begin{cases} -i \log z & \text{, on } \Omega \\ \alpha & \text{, on } B(0, r) \end{cases}$$

where $\alpha < \theta_1$ is a fixed real number. Then h is holomorphic on $\Omega \cup B(0, r)$. By the Holomorphic Functional Calculus we have

$$\mathrm{Sp}\, h(f(\lambda)) \subset \{-i \log z : z \in \mathrm{Sp}\, f(\lambda) \cap \Omega\} \cup \{\alpha\}.$$

So $u(\lambda) = \max\{\mathrm{Re}\, z : z \in \mathrm{Sp}\, h(f(\lambda))\}$. Then we apply Corollary 3.4.9 to $h \circ f$. For $-v$, the proof is similar. $\square$

If x is in a Banach algebra we define the *peripherical spectrum* of x to be the set of $\lambda \in \mathrm{Sp}\, x$ such that $|\lambda| = \rho(x)$.

THEOREM 3.4.11. *Let f be an analytic function from a domain D of $\mathbb{C}$ into a Banach algebra A. Suppose that there exists $\lambda_0 \in D$ such that $\rho(f(\lambda)) \leq \rho(f(\lambda_0))$, for all $\lambda \in D$. Then the peripherical spectrum of $f(\lambda)$ is constant on D.*

PROOF. By the Maximum Principle for subharmonic functions (see Theorem A.1.3) there exists a constant c such that $\rho(f(\lambda)) = c$ on D. If $c = 0$ the result is obvious. So suppose $c > 0$ and that there exist $\lambda_1, \lambda_2 \in D$ and $z \in \mathbb{C}$ such that $|z| = c$, $z \in \mathrm{Sp}\, f(\lambda_1)$, and $z \notin \mathrm{Sp}\, f(\lambda_2)$. Let $a > 0$. Then $g(\lambda) = f(\lambda) + az\mathbf{1}$ is analytic and so, by Theorem 3.4.7, $\lambda \mapsto \rho(f(\lambda) + az\mathbf{1})$ is subharmonic on D. We have

$$\mathrm{Sp}\, g(\lambda) \subset \overline{B}(az, c) \subset \overline{B}(0, (a+1)c).$$

Moreover these two disks are tangent at the point $(a+1)z$. Consequently $\rho(g(\lambda_2)) < (a+1)c$ because $(a+1)z \notin \mathrm{Sp}\, g(\lambda_2)$ and $\mathrm{Sp}\, g(\lambda_2) \subset \overline{B}(az, c)$. But $\rho(g(\lambda)) \leq (a+1)c = \rho(g(\lambda_1))$, for every $\lambda \in \mathbb{C}$. So by the Maximum Principle, $\rho(g(\lambda)) = (a+1)c$ and this is a contradiction. Consequently $\{z : z \in \mathrm{Sp}\, f(\lambda), |z| = \rho(f(\lambda))\}$ is constant. $\square$

COROLLARY 3.4.12. *Let f be an analytic function from a domain D of $\mathbb{C}$ into a Banach algebra A. Suppose that $\mathrm{Sp}\, f(\lambda) \subset \mathbb{R}$ for all $\lambda \in D$. Then $\mathrm{Sp}\, f(\lambda)$ is constant on D.*

PROOF. First we prove that it is locally constant. Let $\lambda_0 \in D$. Replacing $f(\lambda)$ by $\alpha f(\lambda) + \beta 1$, with appropriate $\alpha, \beta > 0$, we may suppose that $\mathrm{Sp}\, f(\lambda_0) \subset\,]0, 2\pi[$. By upper semicontinuity there exists $\delta > 0$ such that $|\lambda - \lambda_0| < \delta$ implies $\mathrm{Sp}\, f(\lambda) \subset\,]0, 2\pi[$. By Theorem 3.3.3, $g(\lambda) = e^{if(\lambda)}$ is defined for $|\lambda - \lambda_0| < \delta$ and $\mathrm{Sp}\, g(\lambda)$ is included in the unit circle. So by Theorem 3.4.11, $\mathrm{Sp}\, g(\lambda) = \exp(i\, \mathrm{Sp}\, f(\lambda))$ is constant for $|\lambda - \lambda_0| < \delta$. But $z \mapsto e^{iz}$ is one-to-one on $]0, 2\pi[$, so $\mathrm{Sp}\, f(\lambda)$ is constant for $|\lambda - \lambda_0| < \delta$. Now let $E = \{\lambda : \lambda \in D, \mathrm{Sp}\, f(\lambda) = \mathrm{Sp}\, f(\lambda_0)\}$. The previous argument shows that E is open and closed in D. So $E = D$. $\square$

EXAMPLE. Let $x, y \in A$. Suppose that $\mathrm{Sp}(x + \lambda y) \subset \mathbf{R}$ for all $\lambda \in \mathbf{C}$. Then by the previous result we have

$$\mathrm{Sp}(x + \lambda y) = \mathrm{Sp}\, x\,, \quad \text{for all } \lambda \in \mathbf{C}\,.$$

Dividing by λ we get

$$\mathrm{Sp}(\mu x + y) = \mu\, \mathrm{Sp}\, x\,, \quad \text{for all } \mu \neq 0\,,$$

so

$$\rho(\mu x + y) = |\mu|\rho(x)\,, \quad \text{for all } \mu \neq 0.$$

But $\mu \mapsto \rho(\mu x + y)$ is subharmonic. So we get

$$\rho(y) = \limsup_{\substack{\mu \to 0 \\ \mu \neq 0}} \rho(\mu x + y) = 0 \text{ (see Theorem A.1.2)}.$$

THEOREM 3.4.13 (SPECTRAL MAXIMUM PRINCIPLE). *Let f be an analytic function from a domain D of $\mathbf{C}$ into a Banach algebra A. Suppose that there exists $\lambda_0 \in D$ such that $\mathrm{Sp}\, f(\lambda) \subset \mathrm{Sp}\, f(\lambda_0)$, for all $\lambda \in D$. Then $\partial\, \mathrm{Sp}\, f(\lambda_0) \subset \partial\, \mathrm{Sp}\, f(\lambda)$ and $\mathrm{Sp}\, f(\lambda_0)\hat{\ } = \mathrm{Sp}\, f(\lambda)\hat{\ }$, for all $\lambda \in D$. In particular if $\mathrm{Sp}\, f(\lambda_0)$ has no interior points or if $\mathrm{Sp}\, f(\lambda)$ does not separate the plane for all $\lambda \in D$, then $\mathrm{Sp}\, f(\lambda)$ is constant on D.*

PROOF. Suppose that $z_0 \in \partial\, \mathrm{Sp}\, f(\lambda_0)$ and $z_0 \notin \partial\, \mathrm{Sp}\, f(\lambda_1)$ for some $\lambda_1 \in D$. Of course z_0 is not an interior point of $\mathrm{Sp}\, f(\lambda_1)$ because in that case it would be interior to $\mathrm{Sp}\, f(\lambda_0)$. So $z_0 \notin \mathrm{Sp}\, f(\lambda_1)$, and hence there exists $r > 0$ such that $\overline{B}(z_0, r) \cap \mathrm{Sp}\, f(\lambda_1) = \emptyset$. Since $z_0 \in \partial\, \mathrm{Sp}\, f(\lambda_0)$ there exists $z_1 \notin \mathrm{Sp}\, f(\lambda_0)$ such that $|z_1 - z_0| < r/3$. Then

$$\mathrm{dist}(z_1, \mathrm{Sp}\, f(\lambda_0)) < r/3 \quad \text{and} \quad \mathrm{dist}(z_1, \mathrm{Sp}\, f(\lambda_1)) > 2r/3\,.$$

But by hypothesis $\mathrm{dist}(z_1, \mathrm{Sp}\, f(\lambda)) \geq \mathrm{dist}(z_1, \mathrm{Sp}\, f(\lambda_0))$. So by Corollary 3.4.8 and the Maximum Principle for subharmonic functions we get $\mathrm{dist}(z_1, \mathrm{Sp}\, f(\lambda)) = \mathrm{dist}(z_1, \mathrm{Sp}\, f(\lambda_0))$ on D. So we have a contradiction at $\lambda = \lambda_1$. Hence $\partial\,\mathrm{Sp}\, f(\lambda_0) \subset \partial\,\mathrm{Sp}\, f(\lambda)$ for all $\lambda \in D$. Let $U(\lambda)$ be the unbounded component of $\mathbb{C}\backslash \mathrm{Sp}\, f(\lambda)$. Then we have $U(\lambda_0) \subset U(\lambda)$. Now suppose $U(\lambda_0) \neq U(\lambda)$. Then there exists $z \in U(\lambda)$ such that $z \in \mathrm{Sp}\, f(\lambda_0)\hat{\ }$. Let z be connected to infinity by an arc Γ included in $U(\lambda)$. Let z_0 be the supremum on Γ, for the order defined by the parametrization, of the points of $\mathrm{Sp}\, f(\lambda_0)$. Then $z_0 \in \mathrm{Sp}\, f(\lambda_0)$ and it cannot be interior, so $z_0 \in \partial\,\mathrm{Sp}\, f(\lambda_0) \subset \partial\,\mathrm{Sp}\, f(\lambda)$. But this is a contradiction because $U(\lambda) \cap \partial\,\mathrm{Sp}\, f(\lambda) = \emptyset$. So $U(\lambda) = U(\lambda_0)$ and then $\mathrm{Sp}\, f(\lambda)\hat{\ } = \mathrm{Sp}\, f(\lambda_0)\hat{\ }$ for all $\lambda \in D$. If $\mathrm{Sp}\, f(\lambda_0)$ has no interior points then $\mathrm{Sp}\, f(\lambda_0) = \partial\,\mathrm{Sp}\, f(\lambda_0) \subset \partial\,\mathrm{Sp}\, f(\lambda) \subset \mathrm{Sp}\, f(\lambda) \subset \mathrm{Sp}\, f(\lambda_0)$; if $\mathrm{Sp}\, f(\lambda)$ does not separate the plane then $\mathrm{Sp}\, f(\lambda) = \mathrm{Sp}\, f(\lambda)\hat{\ }$ and so the proof is complete. $\square$

THEOREM 3.4.14 (LIOUVILLE'S SPECTRAL THEOREM). *Let f be an analytic function from $\mathbb{C}$ into a Banach algebra A. Suppose there exists a bounded set C such that $\mathrm{Sp}\, f(\lambda) \subset C$ for all $\lambda \in \mathbb{C}$. Then $\mathrm{Sp}\, f(\lambda)\hat{\ }$ is constant on $\mathbb{C}$.*

PROOF.   The set $E = \overline{\bigcup_{\lambda \in \mathbb{C}} \mathrm{Sp}\, f(\lambda)}$ is compact. Let $z_0 \notin E$ and $\epsilon > 0$ be such that $B(z_0, \epsilon) \cap E = \emptyset$. Then, by Corollary 3.4.8, $-\log \mathrm{dist}(z_0, \mathrm{Sp}\, f(\lambda))$ is subharmonic on $\mathbb{C}$ and smaller than $-\log \epsilon$. So by Liouville's theorem for subharmonic functions, we conclude that $\mathrm{dist}(z_0, \mathrm{Sp}\, f(\lambda))$ is constant on $\mathbb{C}$. Let $z_1 \in \partial E$ and suppose there exists $\lambda_1 \in \mathbb{C}$ such that $z_1 \notin \partial\,\mathrm{Sp}\, f(\lambda_1)$. Then $z_1 \notin \mathrm{Sp}\, f(\lambda_1)$ because $\mathrm{Sp}\, f(\lambda_1) \subset E$ and $z_1 \in \partial E$. There exists $r > 0$ such that $B(z_1, r) \cap \mathrm{Sp}\, f(\lambda_1) = \emptyset$. Let λ_2 be such that

$$B(z_1, r/5) \cap \mathrm{Sp}\, f(\lambda_2) \neq \emptyset$$

and let $z_2 \in B(z_1, r/5)\backslash E \neq \emptyset$. We have

$$\mathrm{dist}(z_2, \mathrm{Sp}\, f(\lambda_2)) \leq 2r/5 \quad \text{and} \quad \mathrm{dist}(z_2, \mathrm{Sp}\, f(\lambda_1)) \geq 4r/5.$$

So if we apply the first part to $z_0 = z_2$ we get a contradiction. Consequently $\partial E \subset \partial\,\mathrm{Sp}\, f(\lambda)$ which implies $E\hat{\ } \subset \mathrm{Sp}\, f(\lambda)\hat{\ }$, but the converse is true, so $\mathrm{Sp}\, f(\lambda)\hat{\ } = E\hat{\ }$. Hence $\mathrm{Sp}\, f(\lambda)\hat{\ }$ is constant. $\square$

REMARK 1.   With that hypothesis it is false in general that $\mathrm{Sp}\, f(\lambda)$ is constant. For instance, on $l^2(\mathbb{Z})$ with the orthonormal basis $(e_n)_{n\in\mathbb{Z}}$, we consider the two weighted shifts

$$ae_n = \begin{cases} 0 & \text{, if } n = -1 \\ e_{n+1} & \text{, if } n \neq -1 \end{cases} \qquad be_n = \begin{cases} e_0 & \text{, if } n = -1 \\ 0 & \text{, if } n \neq -1. \end{cases}$$

For $\lambda \in \mathbf{C}$ we then have

$$(a + \lambda b)e_n = \begin{cases} \lambda e_0 & , \text{if } n = -1 \\ e_{n+1} & , \text{if } n \neq -1. \end{cases}$$

By Problem 85 of [3] we can deduce that $\operatorname{Sp} a$ is the closed unit disk and that $\operatorname{Sp}(a + \lambda b)$ is the unit circle for $\lambda \neq 0$. So $\operatorname{Sp}(a + \lambda b)$ is bounded but not constant. More easily, we can prove that $\operatorname{Sp} a$ is included in the closed unit disk, that $0 \in \operatorname{Sp} a$, and that $\operatorname{Sp}(a+\lambda b)$ is included in the unit circle for $\lambda \neq 0$. It is obvious that $0 \in \operatorname{Sp} a$ because $ae_{-1} = 0$. For $k \geq 1$ we have $(a + \lambda b)^k e_n = e_{n+k}$ for $n \geq 0$ or $n < -k$, and $(a + \lambda b)^k e_n = \lambda e_{n+k}$ otherwise. So $\|(a + \lambda b)^k\| \leq \max(1, |\lambda|)$, which implies $\rho(a + \lambda b) \leq 1$ by Theorem 3.2.8. For $\lambda \neq 0, a + \lambda b$ is invertible and its inverse satisfies

$$(a + \lambda b)^{-1} e_n = \begin{cases} \dfrac{1}{\lambda} e_{-1} & , \text{if } n = 0 \\ e_{n-1} & , \text{if } n \neq 0. \end{cases}$$

A similar argument shows that $\rho((a+\lambda b)^{-1}) \leq 1$. But Corollary 3.2.10 implies that $1 \leq \rho(a + \lambda b)\rho((a + \lambda b)^{-1})$ so $1 = \rho(a + \lambda b) = \rho((a + \lambda b)^{-1})$ and hence $\operatorname{Sp}(a + \lambda b)$ is included in the unit circle for $\lambda \neq 0$.

THEOREM 3.4.15. *Let f be an analytic function from $\mathbf{C}$ into a Banach algebra A. Then either $\operatorname{Sp} f(\lambda)\hat{}$ is constant or $\bigcup_{\lambda \in \mathbf{C}} \operatorname{Sp} f(\lambda)\hat{}$ is dense in $\mathbf{C}$.*

PROOF. Suppose there exist $z_0 \in \mathbf{C}$ and $r > 0$ such that $B(z_0, r) \cap \operatorname{Sp} f(\lambda)\hat{} = \emptyset$, for all $\lambda \in \mathbf{C}$. Then $u(z) = \frac{1}{z - z_0}$ is holomorphic on a neighbourhood of $\operatorname{Sp} f(\lambda)\hat{}$, for all $\lambda \in \mathbf{C}$. It is easy to see that u maps a polynomially convex subset of $\mathbf{C} \backslash \overline{B}(z_0, r)$ onto a polynomially convex set. So, by Theorem 3.3.3, $\operatorname{Sp} u(f(\lambda))\hat{} = u(\operatorname{Sp} f(\lambda)\hat{})$. But $\operatorname{Sp} u(f(\lambda))\hat{} \subset B(0, 1/r)$. So, by Theorem 3.4.14, $\operatorname{Sp} u(f(\lambda))\hat{}$ is constant. Because u is injective, we conclude that $\operatorname{Sp} f(\lambda)\hat{}$ is constant. $\square$

REMARK 2. Theorem 3.4.15 will be greatly improved in Chapter VII. We shall see that either $\operatorname{Sp} f(\lambda)\hat{}$ is constant or $\mathbf{C} \backslash \bigcup_{\lambda \in \mathbf{C}} \operatorname{Sp} f(\lambda)\hat{}$ is a G_δ-set having zero capacity.

COROLLARY 3.4.16. *Let f be an analytic function from $\mathbf{C}$ into a Banach algebra A. Suppose there exists a constant number C such that*

$$\max\{|\operatorname{Re} u - \operatorname{Re} v| : u, v \in \operatorname{Sp} f(\lambda)\} \leq C$$

for all $\lambda \in \mathbf{C}$. *Then there exists an entire function* h *such that*

$$\mathrm{Sp}\, f(\lambda)\hat{} = h(\lambda) - h(0) + \mathrm{Sp}\, f(0)\hat{}.$$

In particular this happens if the diameter of $\mathrm{Sp}\, f(\lambda)$ *is uniformly bounded on* $\mathbf{C}$.

PROOF.   With the notations of Corollary 3.4.9, $u(\lambda)$ and $-v(\lambda)$ are subharmonic. So $u(\lambda) - v(\lambda)$ is subharmonic and bounded, and so is constant. Let α be such that $u(\lambda) = v(\lambda) + \alpha$ for $\lambda \in \mathbf{C}$. Then $-u = -v - \alpha$ is subharmonic on $\mathbf{C}$, and consequently u is harmonic on $\mathbf{C}$. Then there exists h entire such that $u(\lambda) = \mathrm{Re}\, h(\lambda)$ for all $\lambda \in \mathbf{C}$. The function $g(\lambda) = f(\lambda) - h(\lambda)\mathbf{1}$ is analytic and we have

$$\mathrm{Sp}\, g(\lambda) = \mathrm{Sp}\, f(\lambda) - h(\lambda) \subset \{z : \mathrm{Re}\, z \le 0\}.$$

This implies that $\bigcup_{\lambda \in \mathbf{C}} \mathrm{Sp}\, g(\lambda)\hat{}$ cannot be dense in $\mathbf{C}$. So by Theorem 3.4.15, $\mathrm{Sp}\, g(\lambda)\hat{}$ is constant. Then we have

$$\mathrm{Sp}\, f(\lambda)\hat{} = h(\lambda) - h(0) + \mathrm{Sp}\, f(0)\hat{}. \quad \blacksquare$$

THEOREM 3.4.17.   *Let* f *be an analytic function from a domain* D *of* $\mathbf{C}$ *into a Banach algebra* A. *Suppose that* $\mathrm{Sp}\, f(\lambda) = \{0, \alpha(\lambda)\}$ *for all* $\lambda \in D$, *where* α *is a mapping from* D *into* $\mathbf{C}$. *Then* α *is holomorphic on* D.

PROOF.   By Corollary 3.4.5, α is continuous on D. Let D' be the open subset of D where $\alpha(\lambda) \ne 0$. If D' is empty there is nothing to prove. So suppose D' non-empty. By Radó's extension theorem (see [7], p. 280) it is enough to prove that α is locally holomorphic on D'. Let $\lambda_0 \in D'$. There exist $\delta > 0$ such that for $|\lambda - \lambda_0| < \delta$, we are in the situation of Corollary 3.4.10, in which case $u = v$. Consequently u, v are harmonic on $B(\lambda_0, \delta)$ and there exists k holomorphic on that disk such that $u(\lambda) = \arg \alpha(\lambda) = \mathrm{Im}\, k(\lambda)$. Taking $g(\lambda) = e^{-k(\lambda)} f(\lambda)$ we have $\mathrm{Sp}\, g(\lambda) \subset \mathbf{R}$ for $|\lambda - \lambda_0| < \delta$. By Corollary 3.4.12, $\mathrm{Sp}\, g(\lambda)$ is constant on $B(\lambda_0, \delta)$. So $\alpha(\lambda) = C e^{k(\lambda)}$, for some C, is locally holomorphic. $\blacksquare$

COROLLARY 3.4.18.   *Let* f *be an analytic function from a domain* D *of* $\mathbf{C}$ *into a Banach algebra* A. *Suppose that* $\mathrm{Sp}\, f(\lambda) = \{\alpha(\lambda)\}$ *for all* $\lambda \in D$, *where* α *is a mapping from* D *into* $\mathbf{C}$. *Then* α *is holomorphic on* D.

PROOF.   The proof is almost identical to the previous one. $\blacksquare$

THEOREM 3.4.19. *Let f be an analytic function from a domain D of $\mathbb{C}$ into a Banach algebra A. Suppose that $\operatorname{Sp} f(\lambda)$ lies on a vertical segment for all $\lambda \in D$. Then there exist a holomorphic function h on D and a fixed compact subset K of $\mathbb{R}$ such that $\operatorname{Sp} f(\lambda) = h(\lambda) + iK$, for all $\lambda \in D$.*

PROOF. With the notations of Corollary 3.4.9 we have $u(\lambda) = v(\lambda)$ for $\lambda \in D$. So u, v are harmonic on D. Let us denote by $h(\lambda)$ the element of $\operatorname{Sp} f(\lambda)$ with the smallest imaginary part. Fix $\lambda_0 \in D$ and $\delta > 0$ such that $\overline{B}(\lambda_0, \delta) \subset D$. On $B(\lambda_0, \delta)$ there is a holomorphic function k such that $u(\lambda) = \operatorname{Re} k(\lambda)$. Taking $g_1(\lambda) = -i(f(\lambda) - k(\lambda)\mathbf{1})$ on $B(\lambda_0, \delta)$, we have $\operatorname{Sp} g_1(\lambda) \subset \mathbb{R}$. So, by Corollary 3.4.12, $\operatorname{Sp} g_1(\lambda)$ is constant on $B(\lambda_0, \delta)$. This implies in particular that h is holomorphic on $B(\lambda_0, \delta)$. Then h is holomorphic on all D. Once more, arguing with $g_2(\lambda) = -i(f(\lambda) - h(\lambda)\mathbf{1})$ on D and applying Corollary 3.4.12, we obtain the result. $\square$

Let f be an analytic function from $D \subset \mathbb{C}$ into $\mathfrak{LC}(X)$ and let $\lambda_0 \in D$, $\alpha_0 \in \operatorname{Sp} f(\lambda_0)$ with $\alpha_0 \neq 0$. To simplify suppose that α_0 is an eigenvalue with multiplicity one, or equivalently that the projection asssociated to $\mathcal{N}(f(\lambda_0) - \alpha_0 I)$ has rank one. Then there exist $r, \delta > 0$ such that $|\lambda - \lambda_0| < \delta$ implies that $\operatorname{Sp} f(\lambda) \cap B(\alpha_0, r)$ contains only one eigenvalue $\alpha(\lambda)$. What can be said about this function α? In this particular case it is known that α is holomorphic on $B(\lambda_0, \delta)$. The classical proof depends strongly on the fact that $f(\lambda) \in \mathfrak{LC}(X)$: see for instance the book by I.C. Gohberg and M.G. Krejn, *Introduction à la théorie des opérateurs linéaires non auto-adjoints dans un espace hilbertien*, Paris, 1971, Chapter II.

In the next theorem we shall see that this result is true in general.

THEOREM 3.4.20 (HOLOMORPHIC VARIATION OF ISOLATED SPECTRAL VALUES). *Let f be an analytic function from a domain D of $\mathbb{C}$ into a Banach algebra A. Suppose there exist $\lambda_0 \in D$, $\alpha_0 \in \operatorname{Sp} f(\lambda_0)$ and $r, \delta > 0$ such that $|\lambda - \lambda_0| < \delta$ implies that $\lambda \in D$ and that $\operatorname{Sp} f(\lambda) \cap B(\alpha_0, r)$ contains only one point $\alpha(\lambda)$. Then α is holomorphic on a neighbourhood of λ_0.*

PROOF. If $\operatorname{Sp} f(\lambda_0) = \{\alpha_0\}$ then, by Theorem 3.4.2, we may suppose without loss of generality that $\operatorname{Sp} f(\lambda) = \{\alpha(\lambda)\}$ for $|\lambda - \lambda_0| < \delta$. Then the result follows from Corollary 3.4.18. Suppose now that $\operatorname{Sp} f(\lambda_0)$ is larger than $\{\alpha_0\}$. For the same reasons we may suppose that $|\lambda - \lambda_0| < \delta$ implies $\operatorname{Sp} f(\lambda) \cap \partial B(\alpha_0, r) = \emptyset$, and in particular that $\operatorname{Sp} f(\lambda_0) \cap \{z : |z - \alpha_0| > r\} \neq \emptyset$. So by Theorem 3.4.4, without loss of generality, we may suppose that for $|\lambda - \lambda_0| < \delta$, $\operatorname{Sp} f(\lambda)$ is the union of $\alpha(\lambda) \in B(\alpha_0, r)$ and a non-empty set included in $\{z : |z - \alpha_0| > r\}$. Let h be the holomorphic function defined by $h(z) = z$ on $B(\alpha_0, r)$ and $h(z) = 0$ on

$\{z: |z - \alpha_0| > r\}$. By Theorem 3.3.3, $h(f(\lambda))$ is defined for $|\lambda - \lambda_0| < \delta$ and we have

$$\mathrm{Sp}\, h(f(\alpha)) = \{0, \alpha(\lambda)\}.$$

Then Theorem 3.4.17 implies the result. $\square$

REMARK 3. If α_0 is isolated in $\mathrm{Sp}\, f(\lambda_0)$, it is not true in general that $\mathrm{Sp}\, f(\lambda)$ will have an isolated element near α_0, for λ near λ_0. Theorem 3.4.4 only says that there will be a small component of $\mathrm{Sp}\, f(\lambda)$ near α_0. The previous theorem asserts that it will vary holomorphically if it contains only one point. This theorem will be generalized in Chapter VII (Theorem 7.1.6).

Even if the spectrum function is not continuous it has some weak continuity properties, corresponding to what is called the *fine topology* in potential theory, that is the smallest topology for which all subharmonic functions are continuous.

LEMMA 3.4.21. *Let $\phi_1, \ldots, \phi_n$ be upper semicontinuous functions on an open subset D of $\mathbb{C}$. Let $E \subset D$ and $\lambda_0 \in D \cap \overline{E}$. If $\phi_1(\lambda_0) + \cdots + \phi_n(\lambda_0) = \limsup\{\phi_1(\lambda) + \cdots + \phi_n(\lambda): \lambda \to \lambda_0, \lambda \neq \lambda_0, \lambda \in E\}$, then there exists a sequence (λ_k) converging to λ_0 such that $\lambda_k \neq \lambda_0$, $\lambda_k \in E$, and $\phi_1(\lambda_0) = \lim_{k \to \infty} \phi_i(\lambda_k)$, for $i = 1, \ldots, n$.*

PROOF. Let $\psi(\lambda) = \phi_1(\lambda) + \cdots + \phi_n(\lambda)$. By hypothesis there exists a sequence (μ_k) converging to λ_0, with $\mu_k \neq \lambda_0$ and $\mu_k \in E$, such that $\psi(\lambda_0) = \lim_{k \to \infty} \psi(\mu_k)$. If $n = 1$ the lemma is obvious. Supposing that the lemma is true for $n - 1$ functions, we prove it for n. If $\phi_1(\lambda_0) = \limsup_{k \to \infty} \phi_1(\mu_k)$, then (μ_k) contains a subsequence, which we denote in the same way for convenience, such that $\lim_{k \to \infty} \phi_1(\mu_k) = \phi_1(\lambda_0)$. Then $\sum_{i=2}^{n} \phi_i(\mu_k)$ converges to $\sum_{i=2}^{n} \phi_i(\lambda_0)$ so, by induction hypothesis, there exists a subsequence (μ_k) such that $\lim_{k \to \infty} \phi_i(\mu_k) = \phi_i(\lambda_0)$ for $i = 2, \ldots, n$. But it is also true for $i = 1$, so in this case the proof is complete. If $L = \limsup_{k \to \infty} \phi_1(\mu_k) < \phi_1(\lambda_0)$, then there exists a subsequence (μ_{k_ℓ}) such that $L = \lim_{k \to \infty} \phi_1(\mu_{k_\ell})$, so $\sum_{i=2}^{n} \phi_i(\mu_{k_\ell})$ converges to $\psi(\lambda_0) - L = \phi_1(\lambda_0) - L + \phi_2(\lambda_0) + \cdots + \phi_n(\lambda_0) > \sum_{i=2}^{n} \phi_i(\lambda_0)$. This is a contradiction because $\phi_2 + \cdots + \phi_n$ is upper semicontinuous. $\square$

For the definition of *non-thin sets at a point* see the Appendix.

THEOREM 3.4.22 (WEAK LOWER SEMICONTINUITY OF THE BOUNDARY OF THE SPECTRUM). *Let f be an analytic function from a domain D of $\mathbb{C}$ into a Banach* algebra A. Suppose that $E \subset D \setminus \{\lambda_0\}$ is non-thin at $\lambda_0 \in D \cap \overline{E}$. Then there exists a sequence (μ_k) converging to λ_0, such that $\mu_k \in E$ and

$$\partial \operatorname{Sp} f(\lambda_0) \subset \partial \operatorname{Sp} f(\mu_k) + B(0, 1/k), \text{ for } k \geq 1.$$

PROOF.   Let $\epsilon > 0$ be given and let $B(\xi_1, \epsilon/2), \ldots, B(\xi_n, \epsilon/2)$ be a finite covering of $\partial \operatorname{Sp} f(\lambda_0)$, with $\xi_1, \ldots, \xi_n \in \partial \operatorname{Sp} f(\lambda_0)$. We choose $\eta_1, \ldots, \eta_n \notin \operatorname{Sp} f(\lambda_0)$ such that $|\xi_i - \eta_i| < \epsilon/8$, for $i = 1, \ldots, n$. By Theorem 3.4.2 there exists $r > 0$ such that $\overline{B}(\lambda_0, r) \subset D$ and such that $u_i(z) = 1/(z - \eta_i)$ is holomorphic on a neighbourhood of $\operatorname{Sp} f(\lambda)$ for $|\lambda - \lambda_0| < r$. By Theorem 3.3.3 and 3.4.7, $\phi_i(\lambda) = \rho(u_i(f(\lambda))) > 0$ is subharmonic for $|\lambda - \lambda_0| < r$ and $i = 1, \ldots, n$. Because E is non-thin at λ_0 we have

$$\sum_{i=1}^{n} \phi_i(\lambda_0) = \limsup \left\{ \sum_{i=1}^{n} \phi_i(\lambda) : \lambda \to \lambda_0, \lambda \in E \right\}.$$

So, by Lemma 3.4.21, there exists a sequence (λ_k) converging to λ_0 such that $\lambda_k \in E$ and $\phi_i(\lambda_0) = \lim_{k \to \infty} \phi_i(\lambda_k)$, for $i = 1, \ldots, n$. In particular there exists $\mu(\epsilon) \in E$ such that $|\mu(\epsilon) - \lambda_0| < r$ and $\phi(\mu(\epsilon)) \geq \phi_i(\lambda_0)/2$, for $i = 1, \ldots, n$. By Theorem 3.3.5, we have

$$\operatorname{dist}(\eta_i, \operatorname{Sp} f(\mu(\epsilon))) = \frac{1}{\phi_i(\mu(\epsilon))} \leq \frac{2}{\phi_i(\lambda_0)}$$
$$= 2 \operatorname{dist}(\eta_i, \operatorname{Sp} f(\lambda_0)) \leq 2|\eta_i - \xi_i| < \epsilon/4.$$

Consequently $\overline{B}(\eta_i, \epsilon/4)$ meets $\partial \operatorname{Sp} f(\mu(\epsilon))$. But $\overline{B}(\xi_i, \epsilon/2) \supset \overline{B}(\eta_i, \epsilon/4)$, so $\overline{B}(\xi_i, \epsilon/2)$ meets $\partial \operatorname{Sp} f(\mu(\epsilon))$ for all $i = 1, \ldots, n$. If $\xi \in \partial \operatorname{Sp} f(\lambda_0)$, there exists some ξ_i such that $|\xi - \xi_i| < \epsilon/2$, so there exists some $\zeta_\epsilon \in \partial \operatorname{Sp} f(\mu(\epsilon))$ such that $|\xi - \zeta_\epsilon| < \epsilon$. Consequently $\partial \operatorname{Sp} f(\lambda_0) \subset \partial \operatorname{Sp} f(\mu(\epsilon)) + B(0, \epsilon)$. $\square$

In particular, this theorem can be applied if E is a subdomain U of D and $\lambda_0 \in D$ is a boundary point of U, and also if E is a Jordan arc included in D ending at $\lambda_0 \in D$ (see Theorem A.1.17 and Theorem A.1.18).

COROLLARY 3.4.23.   With the hypotheses of Theorem 3.4.22, suppose that there exists a closed subset F of $\mathbb{C}$ such that $\partial \operatorname{Sp} f(\lambda) \subset F$ for all $\lambda \in E$. Then $\partial \operatorname{Sp} f(\lambda_0) \subset F$. If moreover F has no interior point and does not separate the plane then $\operatorname{Sp} f(\lambda_0) \subset F$.

PROOF.   By Theorem 3.4.22 we have $\partial \operatorname{Sp} f(\lambda_0) \subset F + B(0, 1/k)$ for all $k \geq 1$, so $\partial \operatorname{Sp} f(\lambda_0) \subset F$. If the interior of $\operatorname{Sp} f(\lambda_0)$ is empty the conclusion is obvious. So suppose this interior non-empty. It cannot be included in F since F has no interior points and so there exists z_0 interior to $\mathrm{Sp}\, f(\lambda_0)$ with $z_0 \in \mathbb{C}\backslash F$. But z_0 can be connected to infinity by an arc Γ included in $\mathbb{C}\backslash F$, which is a contradiction because Γ must cross $\partial\, \mathrm{Sp}\, f(\lambda_0) \subset F$. $\square$

This corollary can be applied to $F = \mathbf{R}$ or any Jordan arc.

We now finish this section with two important results. Let A be a Banach algebra and $x \in A$. For an integer $n \geq 1$ we define the n-th spectral diameter of x, denoted $\delta_n(x)$, by

$$\delta_n(x) = \max \left(\prod_{1 \leq i < j \leq n+1} |\lambda_i - \lambda_j| \right)^{\frac{2}{n(n+1)}}$$

for $n+1$ arbitrary points $\lambda_1, \ldots, \lambda_{n+1}$ in $\mathrm{Sp}\, x$. For $n = 1$ it is the classical diameter of $\mathrm{Sp}\, x$ denoted by $\delta(x)$.

THEOREM 3.4.24.   Let f be an analytic function from a domain D of $\mathbb{C}$ into a Banach algebra A. Then for arbitrary $n \geq 1$ the functions $\lambda \mapsto \delta_n(f(\lambda))$ and $\lambda \mapsto \log \delta_n(f(\lambda))$ are subharmonic on D.

PROOF.   Suppose that $n = 1$. Let $x \in A$ and $|\alpha| = 1$. By Corollary 3.4.9 the length of the projection of $\mathrm{Sp}\, x$ on the line $\{t\overline{\alpha} : t \in \mathbf{R}\}$ is given by $\log \rho(e^{\alpha x}) + \log \rho(e^{-\alpha x})$. Consequently

$$\delta(x) = \max_{|\alpha|=1}(\log \rho(e^{\alpha x}) + \log \rho(e^{-\alpha x})).$$

By Theorem 3.4.7, $\lambda \mapsto \log \rho(e^{\alpha f(\lambda)}) + \log \rho(e^{-\alpha f(\lambda)})$ is subharmonic, consequently $\lambda \mapsto \delta(f(\lambda))$ satisfies the mean inequality. But by Theorem 3.4.2, it is upper semi-continuous. So $\lambda \mapsto \delta(f(\lambda))$ is subharmonic. We have $|e^{p(\lambda)}|\delta(f(\lambda)) = \delta(e^{p(\lambda)} f(\lambda))$ for every polynomial p, and $\lambda \mapsto e^{p(\lambda)} f(\lambda)$ is analytic, so by the first part $|e^{p(\lambda)}|\delta(f(\lambda))$ is subharmonic. Then the Beckenbach-Saks theorem implies that $\lambda \mapsto \log \delta(f(\lambda))$ is subharmonic (see Theorem A.1.8). For $n \geq 2$, the proof is much more complicated. It uses tensor products of Banach algebras and joint spectrum, and was given for the first time in 1982 by Z. Słodkowski. In Chapter VII we shall indicate how this result can be obtained more geometrically (Theorem 7.1.3 and Theorem 7.1.13). $\square$

For the definition of *capacity* see the Appendix.

THEOREM 3.4.25 (SCARCITY OF ELEMENTS WITH FINITE SPECTRUM). *Let f be an analytic function from a domain D of $\mathbb{C}$ into a Banach algebra A. Then either the set of $\lambda \in D$ such that $\operatorname{Sp} f(\lambda)$ is finite is a Borel set having zero capacity, or there exist an integer $n \geq 1$ and a closed discrete subset E of D such that $\#\operatorname{Sp} f(\lambda) = n$ for $\lambda \in D \backslash E$ and $\#\operatorname{Sp} f(\lambda) < n$ for $\lambda \in E$. In that case the n points of $\operatorname{Sp} f(\lambda)$ are locally holomorphic functions on $D \backslash E$.*

PROOF. Let $F = \{\lambda: \#\operatorname{Sp} f(\lambda) < +\infty\}$ and $F_k = \{\lambda: \#\operatorname{Sp} f(\lambda) \leq k\}$. Then $\lambda \in F_k$ if and only if $\log \delta_k(f(\lambda)) = -\infty$. But, by Theorem 3.4.24, $\log \delta_k(f(\lambda))$ is subharmonic. So, by H. Cartan's theorem (see Theorem A.1.29), either F_k is a G_δ-set having zero capacity for all $k \geq 1$, or $c(F_n) > 0$ for the smallest integer $n \geq 1$ such that $\log \delta_n(f(\lambda)) \equiv -\infty$ on D, in which case we have $F_n = D$. The first case implies that F is a Borel set and that $c(F) = 0$. The second case implies $\#\operatorname{Sp} f(\lambda) \leq n$, for all $\lambda \in D$. Let $E = \{\lambda: \lambda \in D, \#\operatorname{Sp} f(\lambda) < n\}$. We shall now prove that E is closed in D. Let (λ_ℓ) be a sequence of elements of E converging to $\lambda_0 \in D$. If $\lambda_0 \notin E$ then $\operatorname{Sp} f(\lambda_0)$ contains n distinct points $\alpha_1, \ldots, \alpha_n$. We choose $\epsilon > 0$ such that all the disks $B(\alpha_i, \epsilon)$ are disjoint. By Theorem 3.4.4, for ℓ large enough we would have $\operatorname{Sp} f(\lambda) \cap B(\alpha_i, \epsilon) \neq \emptyset$ for $i = 1, \ldots, n$ and consequently $\#\operatorname{Sp} f(\lambda_\ell) = n$, which is a contradiction. So E is closed. Let $\lambda_0 \notin E$. Then $\operatorname{Sp} f(\lambda_0) = \{\alpha_1, \ldots, \alpha_n\}$. The previous argument, with Theorem 3.4.4, implies in fact that $\#(\operatorname{Sp} f(\lambda) \cap B(\alpha_i, \epsilon)) = 1$ for $|\lambda - \lambda_0|$ small enough. Consequently, by Theorem 3.4.20 we have $\operatorname{Sp} f(\lambda) = \{\alpha_1(\lambda), \ldots, \alpha_n(\lambda)\}$, where the α_i are holomorphic on a neighbourhood of λ_0. For every $\lambda \in D \backslash E$ the function $\phi(\lambda) = \prod_{1 \leq i < j \leq n}(\alpha_i(\lambda) - \alpha_j(\lambda))^2$ is well-defined, holomorphic and not identically zero on $D \backslash E$. If we define ϕ on E by $\phi(\lambda) = 0$ it is not difficult to see that ϕ is continuous on D. If $\lambda_1 \notin E$ it is obviously continuous at λ_1, so suppose that $\lambda_1 \in E$. Then $\operatorname{Sp} f(\lambda_1) = \{\beta_1, \ldots, \beta_k\}$ with $k < n$. From the definition of E and H. Cartan's theorem, E has no interior points. Considering disjoint disks $B(\beta_i, \epsilon)$, we conclude that there exists $r > 0$ such that $|\lambda - \lambda_1| < r$, with $\lambda \notin E$, implies that one of the disks contains at least two points of $\operatorname{Sp} f(\lambda)$. Consequently $|\phi(\lambda)| \leq 2\epsilon\delta(f(\lambda))^{\frac{n(n-1)}{2}-1}$. But $\delta(f(\lambda))$ is upper semicontinous at λ_1, so $\lim_{\lambda \to \lambda_1} \phi(\lambda) = 0$. By the Radó's extension theorem (see [7], p. 280), ϕ is holomorphic on all D. Consequently its set of zeros, precisely E, is discrete. $\square$

We immediately give an application of the scarcity theorem to spectral theory.

In the period 1952-1955, F.V. Atkinson, B.Sz.-Nagy and Ju.L. Šmul'jan proved independently the following result: let $\lambda \mapsto f(\lambda)$ be an analytic function from a domain $D \subset \mathbb{C}$ into the algebra of compact operators on a Banach space and let $z \neq 0$, then the set of $\lambda \in D$ such that $z \in \operatorname{Sp} f(\lambda)$ is a closed and discrete subset of D. Their argument was essentially based on the fact that the projections associated to isolated eigenvalues of compact operators have finite rank. If $D = \mathbf{C}$ and $f(\lambda) = \lambda K$ for some fixed compact operator K then this result says nothing more than Theorem 2.2.10. For a general f, B. Sz.-Nagy believed that this result was deeper than Riesz's theorem. Actually it does not depend on the fact that $f(\lambda)$ is compact but only on the geometry of the graph of the multifunction $\lambda \mapsto \mathrm{Sp}\, f(\lambda)$, namely that $\mathrm{Sp}\, f(\lambda)$ has at most 0 as a limit point for all $\lambda \in D$.

THEOREM 3.4.26.   *Let f be an analytic function from a domain $D \subset \mathbf{C}$ into a Banach algebra A. Suppose that for all $\lambda \in D$ the spectrum of $f(\lambda)$ has at most 0 as a limit point. Let $z \neq 0$. Then either the set of $\lambda \in D$ such that $z \in \mathrm{Sp}\, f(\lambda)$ is closed and discrete in D, or $z \in \mathrm{Sp}\, f(\lambda)$ for all $\lambda \in D$.*

PROOF.   Suppose that $z \in \mathrm{Sp}\, f(\lambda_0)$ for some $\lambda_0 \in D$. We shall prove that λ_0 is either isolated or interior in the set $E = \{\lambda : \lambda \in D, z \in \mathrm{Sp}\, f(\lambda)\}$. Because $z \neq 0$ there exists an open disk $\triangle$ centred at z and not containing 0 such that $\overline{\triangle} \cap \mathrm{Sp}\, f(\lambda_0) = \{z\}$. By Theorem 3.4.4 and upper semicontinuity of the spectrum, there exists $r > 0$ such that $|\lambda - \lambda_0| < r$ implies $\triangle \cap \mathrm{Sp}\, f(\lambda) \neq \emptyset$ and $\partial \triangle \cap \mathrm{Sp}\, f(\lambda) = \emptyset$. Let h be the function defined by

$$h(z) = \begin{cases} z & \text{for } z \in \triangle \\ 0 & \text{for } z \notin \overline{\triangle}. \end{cases}$$

By Theorem 3.3.3 and the hypothesis, we have $\#\, \mathrm{Sp}\, h(f(\lambda)) < +\infty$ and $\#\, (\triangle \cap \mathrm{Sp}\, f(\lambda)) = \#\, (\mathrm{Sp}\, h(f(\lambda)) \backslash \{0\})$ for $|\lambda - \lambda_0| < r$. So, by Theorem 3.4.25, there exist an integer $n \geq 1$, a closed discrete subset F of the disk $B(\lambda_0, r)$ and n functions $\alpha_1, \ldots, \alpha_n$, which are holomorphic on $B(\lambda_0, r) \backslash F$, such that

$$\triangle \cap \mathrm{Sp}\, f(\lambda) = \{\alpha_1(\lambda), \ldots, \alpha_n(\lambda)\}, \quad \text{for } \lambda \in B(\lambda_0, r) \backslash F.$$

There exists s such that $0 < s \leq r$ and $B(\lambda_0, s) \cap F \subset \{\lambda_0\}$. Then the functions $\alpha_1, \ldots, \alpha_n$ are holomorphic on $B(\lambda_0, s)$, except perhaps at λ_0. Moreover, by upper semicontinuity of the spectrum we have $\lim_{\lambda \to \lambda_0} \alpha_i(\lambda) = z$ for $i = 1, 2, \ldots, n$. Therefore the α_i's can be extended holomorphically to the whole disk $B(\lambda_0, s)$. It follows that either $\alpha_{i_0}(\lambda) \equiv z$ for some i_0 or there exists t with $0 < t \leq s$ such that $\alpha_i(\lambda) \neq z$ for all $\lambda \in B(\lambda_0, t) \backslash \{\lambda_0\}$ and $i = 1, \ldots, n$. In the first case λ_0 is an interior point of E while in the second case λ_0 is isolated in E. To finish the argument we consider the set E' of all limit points of E in D. By upper semicontinuity of the spectrum, E is closed, so $E' \subset E$. Let $\mu \in E'$. Since μ is not isolated in E it is an interior point of E, hence an interior point of E'. So E' is both closed and open in D. Consequently we have either $E' = \emptyset$ or $E' = D$ and the proof is complete. $\square$

This result can be generalized supposing only that for every λ, $\mathrm{Sp}\, f(\lambda)$ is at most countable, with the conclusion that E is either closed and at most countable, or all D. But the proof is much more complicated and uses the theory of analytic multifunctions (see Chapter VII).

EXERCISE 1.   Let A be a Banach algebra for a norm $\|\cdot\|$ such that $\|1\| \neq 1$. Setting $\||x\|| = \max_{\|y\| \leq 1} \|xy\|$, prove that $\||xy\|| \leq \||x\|| \cdot \||y\||$, $\frac{\|x\|}{\|1\|} \leq \||x\|| \leq \|x\|$ and $\||1\|| = 1$, for all $x, y \in A$.

EXERCISE 2.   Suppose that A is an algebra which is, at the same time, a Banach space for a norm $\|\cdot\|$. Suppose moreover that the left multiplications $x \mapsto xy$ and right multiplications $x \mapsto xy$ are continuous for all $y \in A$. Prove that A is a Banach algebra for a norm $\||\cdot\||$ equivalent to $\|\cdot\|$.

EXERCISE 3.   Given a Banach algebra A prove that there exists a continuous isomorphism of A onto a closed subalgebra of $\mathcal{L}(X)$, for some suitable Banach space X.

EXERCISE 4.   Prove that $\mathcal{LC}(X) + \mathbb{C}I$ is semi-simple.

*EXERCISE 5.   Let H be a Hilbert space.   Prove that the Calkin algebra $\mathcal{L}(H)/\mathcal{LC}(H)$ is semi-simple. Is it still true in the case of a Banach space?

EXERCISE 6.   Let p be a projection of a Banach algebra A. Prove that pAp is a closed semi-simple subalgebra of A, with unit p.

EXERCISE 7.   Let I be a closed two-sided ideal of a Banach algebra A. Prove that $\mathrm{Rad}\, I = I \cap \mathrm{Rad}\, A$.

EXERCISE 8.   Give another proof of Theorem 3.2.8, part (iii).

EXERCISE 9.   Given n elements $x_1, \ldots, x_n$ of a Banach algebra A, suppose that $x_i x_j = 0$, for $i \neq j$. Prove that $\mathrm{Sp}(x_1 + \cdots + x_n) \backslash \{0\} = (\mathrm{Sp}\, x_1 \cup \cdots \cup \mathrm{Sp}\, x_n) \backslash \{0\}$.

EXERCISE 10.   Let A be a Banach algebra without unit. Denote by e the element $(0, 1)$ of $\tilde{A}$. Prove that

$$\frac{1}{3}(\rho(x) + |\lambda|) \leq \rho(x + \lambda e) \leq \rho(x) + |\lambda|,$$

for all $x \in A$ and $\lambda \in \mathbb{C}$, where ρ is the spectral radius in $\tilde{A}$.

EXERCISE 11.   Let x, y be two commuting elements of a Banach algebra A. Prove that $\mathrm{Sp}(x + y) \subset \mathrm{Sp}\, x + \mathrm{Sp}\, y$. Conclude that $\delta(x + y) \leq \delta(x) + \delta(y)$.

*EXERCISE 12.   Give an example of a Banach algebra A and of an element $x \in A$ such that x is a topological divisor of zero and $0 \notin \partial \mathrm{Sp}\, x$.

EXERCISE 13.   Let A be a Banach algebra and let $x \in A$. Suppose that U is an open set containing $\operatorname{Sp} x$, that f is holomorphic on U, that U_1 is an open set containing $f(\operatorname{Sp} x)$ and that g is holomorphic on U_1. Prove that $(g \circ f)(x) = g(f(x))$.

EXERCISE 14.   Let A be a Banach algebra and let $x \in A$. Suppose that $f(x) = 0$, where f is holomorphic on a neighbourhood of $\operatorname{Sp} x$. Prove that x is algebraic.

EXERCISE 15.   Let A be a Banach algebra and let $x, y \in A$. Prove that $e^{x+y} = \lim_{n \to \infty} \left(e^{x/n} e^{y/n} \right)^n$. Is it true in general that $e^{x+y} = e^x \cdot e^y$?

EXERCISE 16.   Let T be a $n \times n$ matrix and let f be holomorphic in a neighbourhood of the spectrum of T. Suppose that $\alpha_1, \ldots, \alpha_n$ are n distinct numbers in the domain of definition of f. Prove that

$$f(T) = \sum_{i=1}^{n} f(\alpha_i) \times \prod_{j \neq i} \frac{T - \alpha_j I}{\alpha_i - \alpha_j}.$$

EXERCISE 17.   Let A be a commutative Banach algebra. Prove that the group of invertible elements of A is either connected or has an infinite number of components. What happens in the non-commutative case? (See V. Paulsen's result).

EXERCISE 18.   Extend Corollary 3.4.12, supposing that $\operatorname{Sp} f(\lambda) \subset \Gamma$ for all $\lambda \in D$, where Γ in an analytic arc of the complex plane.

EXERCISE 19.   Let f be an analytic function from an open set $D \subset \mathbf{C}$ into a Banach algebra A. Suppose that $\overline{\Delta}$ is a closed disk included in D and that $\operatorname{Sp} f(\lambda) \subset \mathbf{R}$, for all $\lambda \in \partial \Delta$. Prove that $\operatorname{Sp} f(\lambda) \subset \mathbf{R}$ and is constant on $\overline{\Delta}$.

EXERCISE 20.   Let u be an element of a Banach algebra A which is a projection modulo the radical. Prove that there exists a projection in A wich is equal to u modulo the radical.

EXERCISE 21.   Let a, b be two elements of a Banach algebra A. Suppose that $\operatorname{Sp}(a + \lambda b)$ contains only one element, denoted by $\alpha(a + \lambda b)$, for all $\lambda \in \mathbf{C}$. Prove that $\alpha(a + \lambda b) = \alpha(a) + \lambda \alpha(b)$, for all $\lambda \in \mathbf{C}$. As a corollary prove that if A is semi-simple and $\operatorname{Sp} x$ contains one element for all $x \in A$ then A is isomorphic to $\mathbf{C}$.

∗EXERCISE 22.   Let f be an analytic function from a domain $D \subset \mathbf{C}$ into a Banach algebra A.

(i) Suppose that for every $\lambda \in D$ the convex hull of the spectrum of $f(\lambda)$ is a linear segment $[\rho(\lambda)\alpha(\lambda), \alpha(\lambda)]$ where $0 < \rho(\lambda) < 1$. Prove that α is holomorphic on D and that ρ is contant on D.

(ii) Suppose that for every $\lambda \in D$ the convex hull of the spectrum of $f(\lambda)$ is $[0, \alpha(\lambda)]$. Prove that α is holomorphic on D.

Chapter IV

REPRESENTATION THEORY

§1. Gelfand Theory for Commutative Banach Algebras

Let A be a Banach algebra. A linear functional χ on A is called a *character* of A if it is multiplicative and not identical to 0 on A. This last condition is equivalent to saying that $\chi(1) = 1$ because $\chi(x) = \chi(x)\chi(1)$. If χ is a character of A it is easy to verify that $\chi(x) \in \mathrm{Sp}(x)$, for all $x \in A$, because $(x - \chi(x)1)y = y(x - \chi(x)1) = 1$ leads to an absurdity. Consequently $|\chi(x)| \leq \rho(x) \leq \|x\|$, so a character is continuous and of norm one.

Many algebras have no characters, for instance $M_n(\mathbf{C})$ for $n \geq 1$, $\mathfrak{L}(H)$, etc. (see Exercise IV.1).

In 1967-1968, using analytic tools, A. Gleason, J.-P. Kahane and W. Żelazko gave a nice characterization of characters. We now give a very elementary proof of this result which is due to M. Roitman and Y. Sternfeld.

THEOREM 4.1.1. *Let A be a Banach algebra. Then $\chi \in A'$ is a character of A if and only if $\chi(x) \in \mathrm{Sp}\, x$, for all $x \in A$.*

PROOF. Suppose that $\chi(x) \in \mathrm{Sp}\, x$, for all $x \in A$. In particular $\chi(1) \in \mathrm{Sp}\, 1$, so $\chi(1) = 1$. Let $p(\lambda) = \chi((\lambda 1 - x)^n)$ for $n \geq 2$. Obviously p is a polynomial of degree n having n roots $\lambda_1, \ldots, \lambda_n$. We have $0 = p(\lambda_i) = \chi((\lambda_i 1 - x)^n) \in \mathrm{Sp}(\lambda_i 1 - x)^n$ and consequently $\lambda_i \in \mathrm{Sp}\, x$ and $|\lambda_i| \leq \rho(x)$. We have

$$p(\lambda) = \lambda^n - n\chi(x)\lambda^{n-1} + \binom{n}{2}\chi(x^2)\lambda^{n-2} + \cdots + (-1)^n\chi(x^n) = \prod_{i=1}^{n}(\lambda - \lambda_i),$$

and consequently

$$\sum_{i=1}^{n} \lambda_i = n\chi(x) , \qquad \sum_{1 \le i < j \le n} \lambda_i \lambda_j = \binom{n}{2} \chi(x^2).$$

However,

$$\left(\sum_{i=1}^{n} \lambda_i\right)^2 = \sum_{i=1}^{n} \lambda_i^2 + 2 \sum_{1 \le i < j \le n} \lambda_i \lambda_j = \sum_{i=1}^{n} \lambda_i^2 + n(n-1)\chi(x^2).$$

So $n^2|\chi(x)^2 - \chi(x^2)| \le n\rho(x)^2 + n|\chi(x^2)|$. This being true for all $n \ge 2$, we conclude that $\chi(x)^2 = \chi(x^2)$ for all $x \in A$. Then we have $\chi((x+y)^2) = \chi(x^2 + y^2 + xy + yx) = (\chi(x) + \chi(y))^2 = \chi(x^2) + \chi(y^2) + 2\chi(x)\chi(y)$, and so

$$\chi(xy + yx) = 2\chi(x)\chi(y) , \quad \text{for all } x, y \in A. \tag{1}$$

Now the identity

$$(ab - ba)^2 + (ab + ba)^2 = 2[a(bab) + (bab)a] \tag{2}$$

implies

$$(\chi(ab - ba))^2 + 4\chi(a)^2\chi(b)^2 = 4\chi(a)\chi(bab). \tag{3}$$

Taking $a = x - \chi(x)1$ and $b = y$ we have $\chi(a) = 0$. Consequently $\chi(ay) = \chi(ya)$ and hence $\chi(xy) = \chi(yx)$. Finally, from (1) we obtain $\chi(xy) = \chi(x)\chi(y)$. $\square$

We now intend to show that A has plenty of characters if A is commutative and that these characters play a very important rôle concerning the representation of the algebra. This important discovery, which has many consequences in spectral theory, in harmonic analysis and in approximation theory, was made by I.M. Gelfand by 1940.

THEOREM 4.1.2 (I.M. GELFAND). *Let A be a commutative Banach algebra. Then we have the following properties:*

(i) *$\chi \mapsto \text{Ker}\,\chi$ defines a bijection from the set of characters of A onto the set of maximal ideals of A,*

(ii) *for every $x \in A$ we have $\text{Sp}\,x = \{\chi(x): \text{for all characters } \chi \text{ of } A\}$.*

PROOF. (i) If χ is a character of A then obviously $\text{Ker}\,\chi$ is an ideal of A. It is maximal because $\text{codim}\,\text{Ker}\,\chi = 1$. Conversely if I is a maximal ideal of A then, by

Corollary 3.2.2, it is closed. Moreover A/I is a commutative algebra with no non-trivial ideals so A/I is a field. By Corollary 3.2.9, there exists an isomorphism ϕ from A/I onto $\mathbf{C}$. Taking χ as the composition of the canonical morphism $A \mapsto A/I$ and ϕ, χ is a character and $\mathrm{Ker}\,\chi = I$. If $\mathrm{Ker}\,\chi_1 = \mathrm{Ker}\,\chi_2$, then since $\chi_1(1) = \chi_2(1) = 1$ we conclude that $\chi_1 = \chi_2$, so that the previous mapping is a bijection.

(ii)   If χ is a character then $x - \chi(x)1$ is not invertible because $(x - \chi(x)1)y = 1$ implies $0 \cdot \chi(y) = 0 = \chi(1) = 1$, a contradiction. Conversely, if $x - \lambda 1$ is not invertible then $(x - \lambda 1)A$ is an ideal which is, by Lemma 3.1.1, contained in some maximal ideal I. By (i), there exists a character χ_0 such that $I = \mathrm{Ker}\,\chi_0$ so $\chi_0((x - \lambda 1)1) = \chi_0(x) - \lambda = 0$, and so (ii) is proved. $\square$

REMARK 1.   In particular, this theorem implies that the set of characters of A, denoted by $\mathfrak{M}(A)$, is not empty. It also implies that in the commutative case, the radical of A coincides with its set of quasi-nilpotent elements. In fact, if we have $\rho(x) = 0$ then $\chi(x) = 0$ for all $\chi \in \mathfrak{M}(A)$, consequently $\chi(xy) = 0$ for all $y \in A$. Hence $\rho(xy) = 0$ and, by Theorem 3.1.3 (iii), $x \in \mathrm{Rad}\,A$.

In some cases it is possible to determine $\mathfrak{M}(A)$ explicitly.

THEOREM 4.1.3.   *Let K be a compact set and $A = C(K)$. Then $\mathfrak{M}(A)$ can be identified with K. In particular, for every closed ideal I of $C(K)$ there exists a closed subset F of K such that $I = \{f : f \in C(K),\ f(x) = 0$ for all $x \in F\}$.*

PROOF.   For each $x \in K$, $f \mapsto f(x)$ is a character of A, denoted by χ_x and called the *evaluation* at x. Since A separates the points of K then $x \mapsto \chi_x$ is an embedding of K into $\mathfrak{M}(A)$. We now prove that every $\chi \in \mathfrak{M}(A)$ is a χ_x for some $x \in K$. Suppose this is false. Let $\chi \neq \chi_x$ for all $x \in K$. Then for every $x \in K$ there exists $f_x \in A$ such that $f_x(x) \neq 0$ and $\chi(f_x) = 0$. The $V_x = \{y : y \in K, f_x(y) \neq 0\}$ form an open covering of K. So there exist $x_1, \ldots, x_n \in K$ such that the $f_i = f_{x_i}$, for $i = 1, \ldots, n$, satisfy $\chi(f_i) = 0$, for $i = 1, \ldots, n$, and such that for every $x \in K$ there exists an i for which $f_i(x) \neq 0$. Let $g = f_1 \bar{f}_1 + \cdots + f_n \bar{f}_n$. Then $\chi(g) = 0$ and $g(x) > 0$ for all $x \in K$. Consequently $g^{-1} \in C(K)$, which is a contradiction as we would then have $\chi(1) = 0$. Hence $x \mapsto \chi_x$ is a bijection from K onto $\mathfrak{M}(A)$. Let $\bar{I} = \{\bar{f} : f \in I\}$. It is easy to verify that $I \cap \bar{I}$ is an ideal of $C(K)$. Let $F = \{x : x \in K, f(x) = 0$, for all $f \in I \cap \bar{I}\}$. Because $f(x) = 0$ is equivalent to $\bar{f}f(x) = 0$ we have $F = \{x : x \in K, f(x) = 0$, for all $f \in I\}$. Let $J = \cap_{x \in F} \mathrm{Ker}\,\chi_x$. Then $I \cap \bar{I} \subset I \subset J$. We shall now prove that $J \subset I \cap \bar{I}$ and the theorem will thus be proved. Let K' be the *contraction* of K, that is the compact topological space obtained from K by the equivalence relation

$$x \equiv y \Leftrightarrow \begin{cases} x = y, & \text{if } x \notin F \\ y \in F, & \text{if } x \in F. \end{cases}$$

Denoting by a the point of K' corresponding to F, J can be identified with the maximal ideal of functions of $C(K')$ vanishing at a. Identifying $I \cap \bar{I}$ with its image in $C(K')$ it is easy to verify that $(I \cap \bar{I}) + \mathbf{C}1$ separates the points of K' and so, by the Stone-Weierstrass theorem, $(I \cap \bar{I}) + \mathbf{C}1 = C(K')$. Hence if $f \in J$ then $f = \lambda 1 + g$ with $g \in I \cap \bar{I}$, so $\lambda 1 = f - g \in J$, and consequently $\lambda = 0$. So the assertion is proved. $\square$

In his famous book *Tauberian Theorems*, N. Wiener proved that a continuous function defined on $\mathbf{R}$, with period 2π, which has an absolutely converging trigonometric series ($f \in W$ in the terminology of Example 3, Chapter III, §1) and such that $f(x) \neq 0$ for $0 \leq x \leq 2\pi$, has also an inverse with an absolutely converging trigonometric series. The original proof is long and complicated. This result was extended by P. Lévy for the composition of f with a function h holomorphic on a neighbourhood of the set of values of f. In 1940, I.M. Gelfand surprised the mathematical world by giving both results a very simple proof. We now see his argument.

THEOREM 4.1.4.   *Let W be the commutative Banach algebra of absolutely converging trigonometric series $\sum_{n=-\infty}^{\infty} a_n e^{int}$, with sum, product and norm defined by $\|(a_n)\| = \sum_{n=-\infty}^{\infty} |a_n|$. Then $\mathfrak{M}(W)$ can be identified with the interval $[0, 2\pi]$.*

PROOF.   If $x \in [0, 2\pi]$ then $\chi_x(f) = \sum_{n=-\infty}^{\infty} a_n e^{inx}$, where $f(t) = \sum_{n=-\infty}^{\infty} a_n e^{int}$, defines a character of W. We now prove the converse, namely that for every character $\chi \in \mathfrak{M}(W)$ there exists $x \in [0, 2\pi]$ such that $\chi = \chi_x$. Applying χ to the two functions e^{it} and e^{-it} we have

$$|\chi(e^{it})| \leq \|e^{it}\| = 1 \quad \text{and} \quad |\chi(e^{-it})| \leq \|e^{-it}\| = 1.$$

But $\chi(e^{it}) \cdot \chi(e^{-it}) = 1$, so $|\chi(e^{it})| = 1$. Let $x \in [0, 2\pi]$ such that $\chi(e^{it}) = e^{ix}$. By continuity of χ we obtain immediately that

$$\chi(f) = \sum_{n=-\infty}^{\infty} a_n e^{inx} = \chi_x(f) , \quad \text{for all } f \in W.$$

So $\chi = \chi_x$. $\square$

COROLLARY 4.1.5 (N. WIENER-P. LÉVY). *Let $f \in W$ and let h be holomorphic on a neighbourhood of the set of values of f . Then $h \circ f \in W$.*

PROOF. By Theorem 4.1.2 (ii) and Theorem 4.1.3 we have $\mathrm{Sp}\, f = \{\chi(f) : \chi \in \mathfrak{M}(W)\} = \{\chi_x(f) : x \in [0, 2\pi]\} = f([0, 2\pi])$. By Theorem 3.3.3, we have $h \circ f = h(f) \in W$. $\square$

In particular, if $f(x) \neq 0$ for $0 \leq x \leq 2\pi$ then $1/z$ is holomorphic on a neighbourhood of the set of values of f, so $1/f \in W$.

We give another application.

THEOREM 4.1.6. *Let K be a compact subset of $\mathbb{C}$ and let $A(K)$ be the Banach algebra of continuous functions on K which are holomophic on the interior of K. Then $\mathfrak{M}(A(K))$ can be identified with K.*

PROOF. The mapping $x \mapsto \chi_x$ is an injective mapping from K into $\mathfrak{M}(A(K))$ because the function $I(z) = z$ is in $A(K)$. Conversely we prove that all $\chi \in \mathfrak{M}(A(K))$ have the form χ_x. We have $x = \chi(I) \in K$, because otherwise the function defined by

$$g(z) = \frac{1}{z - \chi(I)}$$

is in $A(K)$, so that we get $\chi(g \cdot (I - \chi(I)\mathbf{1})) = \chi(\mathbf{1}) = 0$ which is a contradiction. Since $\chi(I) = x$, by continuity and the local series representation of f, we get $\chi(f) = f(x)$, so $\chi = \chi_x$. $\square$

COROLLARY 4.1.7. *Let $f_1, \ldots, f_n \in A(K)$ be such that for all $x \in K$ there exists at least one i such that $f_i(x) \neq 0$. Then there exist $g_1, \ldots, g_n \in A(K)$ such that $f_1 g_1 + \cdots + f_n g_n = 1$.*

PROOF. Let I be the set of all sums $f_1 g_1 + \cdots + f_n g_n$ with $g_1, \ldots, g_n \in A(K)$. If $I \neq A$, then it is contained in some maximal ideal $\mathrm{Ker}\, \chi_x$, and consequently $f_i(x) = 0$ for $i = 1, \ldots, n$ which is a contradiction. So $I = A$, consequently $\mathbf{1}$ is such a sum. $\square$

For some commutative Banach algebras A the set of characters $\mathfrak{M}(A)$ may be extremely complicated and very difficult to determine explicitly. This occurs in the next example.

Let U be an open subset of $\mathbb{C}$. We denote by $H^\infty(U)$ the Banach algebra of functions which are holomorphic and bounded on U. Then the open set U can be embedded in $\mathfrak{M}(A)$ by $x \mapsto \chi_x$ (see K. Hoffman, *Banach Spaces of Analytic Functions*, Englewood Cliffs, 1962). It is a famous problem, called the *Corona problem*, to determine if U is dense in $\mathfrak{M}(A)$ for the Gelfand topology which we define below. This result is true for U simply or finitely connected. The first proof, a complicated one, was given by L. Carleson. Now there is a rather simple proof of this fact due to T. Wolff (see for instance T.W. Gamelin, *Wolff's proof of the Corona Theorem*, Israel J. Math. **37** (1980), pp. 113–119). But the problem remains unsolved for a general open set.

DEFINITION. Let A be a commutative Banach algebra and let $x \in A$. We define the *Gelfand transform* of x, denoted $\hat{x}$, by the formula

$$\hat{x}(\chi) = \chi(x) \quad \text{for } \chi \in \mathfrak{M}(A).$$

Then $\hat{x}$ is a function defined on $\mathfrak{M}(A)$ and $x \mapsto \hat{x}$ is a morphism.

The set $\mathfrak{M}(A)$ is included in the unit ball of the topological dual of A. The *Gelfand topology* on $\mathfrak{M}(A)$ is by definition the restriction of the weak $*$-topology on this set, that is the weakest topology that makes every $\hat{x}$ continuous.

THEOREM 4.1.8 (I.M. GELFAND). *Let $\mathfrak{M}(A)$ be the set of characters of a commutative Banach algebra A. We have the following properties:*

(i) *$\mathfrak{M}(A)$ is compact for the Gelfand topology,*

(ii) *the Gelfand transform is a continuous morphism from A onto a subalgebra $\hat{A}$ of $C(\mathfrak{M}(A))$ whose kernel is $\operatorname{Rad} A$. It is an isomorphism if and only if A is semi-simple,*

(iii) *for each $x \in A$ the range of $\hat{x}$ is the spectrum of x and $\|\hat{x}\|_\infty = \rho(x)$.*

PROOF. (iii) is obvious by Theorem 4.1.2(ii). It is also obvious that $x \mapsto \hat{x}$ is a morphism. It is continuous because $\|\hat{x}\|_\infty = \rho(x) \leq \|x\|$. The last part of (ii) comes from (iii). So we now prove (i). By Theorem 1.1.8, it is sufficient to prove that $\mathfrak{M}(A)$ is weak $*$-closed in the unit ball of A'. Let $f_0 \in A'$ be in the weak $*$-closure of $\mathfrak{M}(A)$. We have only to prove that $f_0(xy) = f_0(x)f_0(y)$, for all $x, y \in A$ and $f_0(1) = 1$. We fix $x, y \in A$ and $\epsilon > 0$. Let $U = \{f : f \in A', |f(z) - f_0(z)| < \epsilon, \text{ for } z = 1, x, y, xy\}$. Then U is a neighbourhood of f_0 for the weak $*$-topology which contains some $\chi \in \mathfrak{M}(A)$. So we have

$$|1 - f_0(1)| = |\chi(1) - f_0(1)| < \epsilon,$$

$$f_0(xy) - f_0(x)f_0(y) = f_0(xy) - \chi(xy) + (\chi(y) - f_0(y))\chi(x) + (\chi(x) - f_0(x))f_0(y),$$

and consequently

$$|f_0(xy) - f_0(x)f_0(y)| \leq \epsilon(1 + \|x\| + |f_0(y)|) , \quad \text{for every } \epsilon > 0.$$

So $\mathfrak{M}(A)$ is weak *-closed in the unit ball of A', and hence weak *-compact. $\square$

COROLLARY 4.1.9. *Let T be a morphism from a commutative Banach algebra A into a semi-simple commutative Banach algebra B. Then T is continuous.*

PROOF. Suppose $\lim x_n = 0$ and $\lim T x_n = a$. By the Closed Graph Theorem it is enough to show that $a = 0$. Let $\chi \in \mathfrak{M}(B)$. Then $\chi \circ T$ is a character of A, and hence it is continuous. So we have

$$\chi(a) = \lim_{n \to \infty} \chi(T x_n) = \lim_{n \to \infty} (\chi \circ T)(x_n) = 0$$

for every $\chi \in \mathfrak{M}(B)$. So $a \in \operatorname{Rad} B = \{0\}$. $\square$

COROLLARY 4.1.10. *On a semi-simple commutative Banach algebra all the Banach algebra norms are equivalent.*

PROOF. Let $\|\cdot\|_1, \|\cdot\|_2$ be two Banach algebra norms on A. We apply the previous corollary to the identity mapping from $(A, \|\cdot\|_1)$ onto $(A, \|\cdot\|_2)$. $\square$

For the definition of Banach algebras with involution see Chapter VI, §1.

COROLLARY 4.1.11. *Every involution on a semi-simple commutative Banach algebra is continuous.*

PROOF. Let us consider the new norm $\||x\|| = \|x^\star\|$. It follows immediately from the definition of an involution that $\|| \cdot \||$ is submultiplicative and $\||1\|| = 1$. If (x_n) is a Cauchy sequence for $\|| \cdot \||$, then $(x_n^\star)$ is a Cauchy sequence for $\| \cdot \|$, and consequently it converges to $a \in A$. Then $\lim_{n \to \infty} \||x_n - a^\star\|| = 0$. So $\|| \cdot \||$ is a Banach algebra norm on A. By Corollary 4.1.10 there exists $C > 0$ such that $\||x\|| = \|x^\star\| \leq C\|x\|$, for $x \in A$, so the involution is continuous. $\square$

The three last corollaries will be improved upon in Chapter V.

COROLLARY 4.1.12. *Let $[a, b]$ be a bounded interval of $\mathbf{R}$. Then the algebra $C^\infty([a, b])$ has no Banach algebra norm.*

PROOF. Suppose that $C^\infty([a, b])$ is a Banach algebra for the norm $\| \cdot \|$. The identity mapping is a morphism from $C^\infty([a, b])$ into $C([a, b])$, so by Corollary 4.1.9 there exists $\alpha > 0$ such that $\|f\|_\infty \leq \alpha\|f\|$, for all $f \in C^\infty([a, b])$. Let $D: f \mapsto f'$ be the linear mapping from $C^\infty([a, b])$ into itself. We now prove that it is continuous. Suppose that $\lim_{n\to\infty} \|f_n\| = 0$ and $\lim_{n\to\infty} \|f'_n - g\| = 0$, for some $g \in C^\infty([a, b])$. By the previous inequality we have $\lim_{n\to\infty} \|f_n\|_\infty = \lim_{n\to\infty} \|f'_n - g\|_\infty = 0$. So $G(t) = \int_a^t g(s)\, ds$ is identically zero on $[a, b]$, and consequently $g = 0$. By the Closed Graph Theorem, D is continuous, hence there exists $\beta > 0$ such that $\|f'\| \leq \beta\|f\|$, for $f \in C^\infty([a, b])$. For $\lambda \in \mathbf{C}$, this second inequality implies that

$$T_\lambda f = f + \frac{\lambda f'}{1!} + \cdots + \frac{\lambda^n f^{(n)}}{n!} + \cdots$$

converges in $C^\infty([a, b])$. By Leibniz's formula for the derivative of products it is easy to verify that T_λ is a morphism from $C^\infty([a, b])$ into itself. If $f \in \operatorname{Rad} C^\infty([a, b])$ then $\chi_x(f) = f(x) = 0$ for all $a \leq x \leq b$. So $C^\infty([a, b])$ is semi-simple. By Corollary 4.1.9 there exists $\gamma > 0$ such that

$$\|T_\lambda f\| \leq \gamma\|f\|, \quad \text{for all } f \in C^\infty([a, b]).$$

The entire function $\lambda \mapsto T_\lambda f$ is bounded so, by Liouville's theorem, $T_\lambda f = f$, and hence $f' = 0$, for all $f \in C^\infty([a, b])$ which is absurd. So the assertion is proved. $\square$

A commutative Banach algebra is called a *function algebra* on a compact set K if it is isometrically isomorphic to a closed subalgebra of $C(K)$ which separates the points of K and contains the constants.

THEOREM 4.1.13. *A Banach algebra A is a function algebra if and only if $\|x^2\| = \|x\|^2$, for all $x \in A$.*

PROOF. The necessity is obvious. So suppose that $\|x^2\| = \|x\|^2$ for all $x \in A$. We first prove that A is commutative. By induction we have $\|x^{2^n}\| = \|x\|^{2^n}$, for $n \geq 0$. So, by Theorem 3.2.8 (iii), $\rho(x) = \|x\|$. Fixing $x, y \in A$, and using Lemma 3.1.2, for all $\lambda \in \mathbf{C}$, we have

$$\|e^{-\lambda y} x e^{\lambda y}\| = \rho(e^{-\lambda y} x e^{\lambda y}) = \rho(x e^{-\lambda y} e^{\lambda y}) = \rho(x).$$

So the analytic function $\lambda \mapsto e^{-\lambda y} x e^{\lambda y}$ is bounded in norm. By Liouville's theorem it is constant, and consequently $xy = yx$ for all $x, y \in A$. By Theorem 4.1.8 we have $\|\hat{x}\|_\infty = \|x\|$, for the Gelfand transform. This implies in particular that $\hat{A}$ is closed in $C(\mathfrak{M}(A))$. But obviously it contains the constants and separates the points of $\mathfrak{M}(A)$. So the result is proved. $\square$

In Theorems 4.1.4 and 4.1.6 we saw that it is easy to characterize the set of characters of some given commutative algebras. We shall prove that this nice situation always occurs when the algebra is finitely generated.

Given K a compact subset of $\mathbf{C}^n$ we recall that the *polynomial convex hull* $\hat{K}$ of K is defined by

$$\hat{K} = \{z: z \in \mathbf{C}^n, |p(z)| \leq \max_{u \in K} |p(u)| \text{ for all polynomials } p\}.$$

Then we say that K is *polynomially convex* if $K = \hat{K}$. If $n = 1$ then $\hat{K}$ is the union of K with its holes. But if $n \geq 2$ then $\hat{K}$ has no topological characterization. Even if K is nicely defined, $\hat{K}$ may be extremely difficult to determine explicitly. By Exercise IV.5, the set of characters of $P(K)$ can be identified with $\hat{K}$. More generally we have:

THEOREM 4.1.14. *Let A be a commutative Banach algebra having n topological generators. Then $\mathfrak{M}(A)$ can be identified with a compact polynomially convex subset of $\mathbf{C}^n$.*

PROOF. Let $x_1, \ldots, x_n$ be topological generators of A. Consequently every element of A is a limit of polynomials in $x_1, \ldots, x_n$. We consider the mapping ϕ from $\mathfrak{M}(A)$ into $\mathbf{C}^n$ defined by $\phi: \chi \mapsto (\chi(x_1), \ldots, \chi(x_n))$.

We prove that it is a homeomorphism from $\mathfrak{M}(A)$ onto $K = \phi(\mathfrak{M}(A))$ and that K is polynomially convex in $\mathbf{C}^n$. By the definition of the Gelfand topology, ϕ is continuous. If $\phi(\chi_1) = \phi(\chi_2)$, then χ_1 and χ_2 coincide on the polynomials in $x_1, \ldots, x_n$ which are dense in A, so $\chi_1 = \chi_2$. But $\mathfrak{M}(A)$ is compact, so ϕ is a homeomorphism from $\mathfrak{M}(A)$ onto K. We now prove that K is polynomially convex. Suppose that $|p(z)| \leq \max_{u \in K} |p(u)|$ for all polynomials p. Then

$$|p(z)| \leq \max_{\chi \in \mathfrak{M}(A)} |p(\chi(x_1), \ldots, \chi(x_n))| = \max_{\chi \in \mathfrak{M}(A)} |\chi(p(x_1, \ldots, x_n))|$$

$$= \rho(p(x_1, \ldots, x_n)) \leq \|p(x_1, \ldots, x_n)\|.$$

Hence $\chi_0(p(x_1, \ldots, x_n)) = p(z)$ defines a character on the algebra of polynomials in $x_1, \ldots, x_n$ which can be extended continuously to A. Since $z = (\chi_0(x_1), \ldots, \chi_0(x_n))$ we have $z \in K$. $\square$

The original proof of Nagasawa's theorem, stated below, uses Choquet's boundary and is rather complicated. We now give a simplified proof based on ideas of G. Niestegge.

Let A be a commutative Banach algebra. We say that a linear functional f on A is a *spectral state* if it satisfies $f(1) = 1$ and $|f(x)| \leq \rho(x)$ for all $x \in A$. The last condition implies that $|f(x)| \leq \|x\|$, so f is continuous. Obviously a character is a spectral state. For every subset E of a vector space, its convex hull is denoted by co E.

LEMMA 4.1.15. *A linear functional f is a spectral state on A if and only if $f(x) \in$ co Sp x for all $x \in A$.*

PROOF. If $f(x) \in$ co Sp x for all $x \in A$ then $f(1) \in$ co Sp $1 = \{1\}$, so $f(1) = 1$, and moreover $|f(x)| \leq \max\{|z|: z \in \text{co Sp } x\} = \rho(x)$. Conversely, for $\lambda \notin$ co Sp x we have $|f(x) - \lambda| = |f(x - \lambda 1)| \leq \rho(x - \lambda 1)$, which means that $f(x)$ is in the disk centred at λ with radius $\rho(x - \lambda 1)$. But co Sp x is the intersection of all such disks $B(\lambda, \rho(x - \lambda 1))$ for $|\lambda| > \rho(x)$, so $f(x) \in$ co Sp x (see Exercise IV.8). $\square$

The set $\mathfrak{F}$ of spectral states is convex, contained in the unit ball of A', and weak*-closed. So, by Theorem 1.1.8 it is weak*-compact. By Theorem 1.1.9 it is the closed convex hull of its extreme points.

LEMMA 4.1.16. *If f is an extreme spectral state then it is a character of A.*

PROOF. Let $\mathfrak{M}$ denote the set of characters of A. Then $\overline{\text{co }\mathfrak{M}} \subset \mathfrak{F}$. Suppose there exists $f_0 \in \mathfrak{F}$ such that $f_0 \notin \overline{\text{co }\mathfrak{M}}$. So, by Corollary 1.1.6, there exists $x_0 \in A$ such that

$$\sup\{\text{Re } f(x_0): f \in \overline{\text{co }\mathfrak{M}}\} < \text{Re } f_0(x_0).$$

Consequently

$$\text{Re } f_0(x_0) > \max\{\text{Re } \chi(x_0): \chi \in \mathfrak{M}\} = \max\{\text{Re } z: z \in \text{Sp } x_0\}$$
$$= \max\{\text{Re } z: z \in \text{co Sp } x_0\}$$

and this gives a contradiction because $f_0(x_0) \in$ co Sp x_0. So co $\mathfrak{M} = \mathfrak{F}$. By Theorem 1.1.12 the set of extreme points of $\mathfrak{F}$ is included in $\mathfrak{M}$. $\square$

THEOREM 4.1.17 (M. NAGASAWA). *Let A and B be commutative and semi-simple Banach algebras. Suppose that T is a linear mapping from A onto B such that $T1 = 1$ and $\rho(Tx) = \rho(x)$ for all $x \in A$. Then T is an isomorphism from A onto B.*

PROOF. Let f be an extreme spectral state on B and let $g = f \circ T$. Then we have $g(1) = 1$ and $|g(x)| = |f(Tx)| \leq \rho(Tx) = \rho(x)$ for all $x \in A$. So g is a spectral state on A. Suppose that $g = \frac{g_1 + g_2}{2}$ where g_1, g_2 are spectral states on A. With the hypotheses, the mapping T is bijective, so for all $y \in B$ we have

$$f(y) = \frac{g_1(T^{-1}y) + g_2(T^{-1}y)}{2}.$$

As was done previously, it is easy to verify that $g_1 \circ T^{-1}$ and $g_2 \circ T^{-1}$ are spectral states on B. Hence, f being extreme, we have $f = g_1 \circ T^{-1} = g_2 \circ T^{-1}$. Consequently $g = g_1 = g_2$ and g is an extreme spectral state of A. By Lemma 4.1.16, g is a character on A. Hence $f(Txy) = f(Tx)f(Ty)$ for all $x, y \in A$. This being true for all extreme spectral states f on B, by Theorem 1.1.9 it is true for all spectral states, and in particular for all characters of B. So $Txy = TxTy$ for all $x, y \in A$, because B is semi-simple. $\square$

LEMMA 4.1.18. *Let K be a compact space. Then f is an extreme point of the closed unit ball of $C(K)$ if and only if $|f(z)| = 1$, for all $z \in K$.*

PROOF. Suppose that $|f(z)| = 1$ for all $z \in K$ and suppose that $f = \frac{u+v}{2}$ where $u, v \in C(K)$ and $\|u\| = \|v\| \le 1$. Taking $u_1 = u/f$, $v_1 = v/f$, we have $1 = \frac{u_1 + v_1}{2}$ where $\|u_1\| = \|v_1\| \le 1$. Let $z \in K$ be arbitrary. Then we have $2 = u_1(z) + v_1(z)$ and $|u_1(z)| \le 1$, $|v_1(z)| \le 1$. Consequently $u_1(z) = v_1(z) = 1$, so $u_1 = v_1 = 1$, and hence f is extreme. Conversely suppose that $\|f\| = 1$ and $|f(z)| < 1$ at some point of K. Then, by continuity, we have the same inequality on a closed set $E \ne K$ having a non-empty interior. Because K is normal, there exists $h \in C(K)$ with support included in E and such that $\sup_{z \in E} |h(z)| < 1 - \sup_{z \in E} |f(z)|$. Now we have

$$f = \frac{(f+h) + (f-h)}{2} \quad \text{and} \quad \|f + h\| = \|f - h\| = \sup_{z \in K \setminus E} |(f \pm h)(z)| = 1,$$

and consequently f is not extremal. $\square$

COROLLARY 4.1.19 (S. BANACH-M. STONE). *Let K_1, K_2 be two compact sets. Suppose that T is a linear isometry from $C(K_1)$ onto $C(K_2)$. Then there exist $u \in C(K_2)$ such that $|u(x)| = 1$ for all $x \in K_2$ and a homeomorphism h from K_2 onto K_1 such that*

$$(Tf)(x) = u(x)f(h(x)) \quad \text{for all } x \in K_2 \text{ and } f \in C(K_1).$$

PROOF. Let $u = T1$. Then $\|u\| = 1$. Suppose that $u = \frac{f+g}{2}$ with $\|f\|, \|g\| \le 1$. Then $1 = \frac{T^{-1}f + T^{-1}g}{2}$ and $\|T^{-1}f\|, \|T^{-1}g\| \le 1$. By Lemma 4.1.18, $1 = T^{-1}f =$

$T^{-1}g$, so $u = f = g$, and hence u is extremal. Consequently $|u(x)| = 1$ for all $x \in K_2$. Let S be the linear mapping defined by $Sf(x) = \frac{Tf(x)}{u(x)}$. We have $\|Sf\| = \|Tf\| = \|f\|$ and $S1 = 1$. Because the norm coincides with the spectral radius on $C(K)$ we conclude, by Theorem 4.1.17, that S is an isomorphism from $C(K_1)$ on $C(K_2)$. Consequently $f \mapsto Sf(x)$ is a character, and so has the form $f \mapsto f(h(x))$ for some point $h(x) \in K_1$. It is now easy to verify that h is bijective and continuous and so is a homeomorphism. $\square$

REMARK 2.   In the hypothesis of Corollary 4.1.19 it is sufficient to suppose that $T0 = 0$, $T(if) = iTf$ and $\|Tf - Tg\| = \|f - g\|$, for all $f, g \in C(K_1)$. Then the linearity of T results from a classical theorem due to S. Mazur and S. Ulam (see S. Banach, *Théorie des opérations linéaires*, New York, 1955, pp. 166–168).

§2. Representation Theory for Non-Commutative Banach Algebras

Let A be a Banach algebra. We say that π is a *representation* of A on a complex vector space X of dimension larger or equal to one if π is a non-trivial morphism from A into the algebra of linear operators on X. If a linear subspace Y of X satisfies $\pi(x)Y \subset Y$ for all $x \in A$, we say that Y is *invariant under* $\pi(x)$. A representation π is said to be *irreducible* if the only linear subspaces of X invariant under $\pi(x)$ are $\{0\}$ and X. This representation is said to be *bounded* if X is a Banach space and if $\pi(x)$ is a bounded linear operator on X for all $x \in A$. Moreover it is said to be *continuous* if it is bounded and if there exists a constant $C > 0$ such that $\|\pi(x)\| \leq C\|x\|$ for all $x \in A$.

For instance if A is commutative then every character χ is a continuous irreducible representation on $X = \mathbb{C}$. If $A = \mathfrak{L}(X)$ for some Banach space X then $\pi(x) = x$ is a continuous irreducible representation on X.

Let L be a maximal left ideal of A (which exists by Lemma 3.1.1). It is closed by Corollary 3.2.2, so $X = A/L$ is a Banach space for the norm $|||\bar{a}||| = \inf_{u \in L} \|a + u\|$. There is a natural representation π of A on X defined by

$$\pi(x)\bar{a} = \overline{xa}.$$

This continuous representation is called the *left regular representation* associated to L. It is irreducible because L is a maximal left ideal. The kernel of the representation is $(L : A) = \{x : x \in A, xA \subset L\}$.

THEOREM 4.2.1. *We have the following properties:*

(i) *for every irreducible representation π of A there exists a maximal left ideal L
such that $\operatorname{Ker} \pi = (L : A)$, consequently $\operatorname{Ker} \pi$ is a closed two-sided ideal of A,*

(ii) *the radical of A is the intersection of the kernels of all continuous irreducible
representations of A,*

(iii) *for every x in A the spectrum of x is the union of all the spectra of the $\pi(x)$ in
the corresponding algebras $\pi(A)$ for all continuous irreducible representations
π.*

PROOF. (i) Let π be an irreducible representation on a complex vector space X
and let $\xi \neq 0$ be in X. Then $F = \{\pi(x)\xi : x \in A\}$ contains ξ and is invariant
under π, so $F = X$. Let $L = \{x : x \in A, \pi(x)\xi = 0\}$. Because $F = X$ we have
$L \neq A$, so L is a left ideal of A. Let J be a proper left ideal of A containing L.
Then either $J = L$ or $\{\pi(x)\xi : x \in J\}$ is different from $\{0\}$ and is invariant under π.
So $\{\pi(x)\xi : x \in J\} = X$. Consequently there exists $e \in J$ such that $\pi(e)\xi = \xi$, so
$xe - x \in L$ for all $x \in A$. Then for any $x \in A$ we have $x = (x - xe) + xe \in L + J \subset J$,
so that $A = J$, which is a contradiction. Consequently L is a maximal left ideal.
Obviously $\operatorname{Ker} \pi \subset (L : A)$. If $x \in (L : A)$ then $xA \subset L$, so $\pi(x)\pi(y)\xi = 0$ for every
$y \in A$. Consequently $\pi(x)X = 0$, hence $\pi(x) = 0$, that is $x \in \operatorname{Ker} \pi$.

(ii) If $x \in \operatorname{Rad} A$ then $xA \subset \operatorname{Rad} A \subset L$ for all maximal left ideals L. Consequently $x \in (L : A)$ for all such maximal left ideals L. By part (i), x is in the
intersection of the kernels of all continuous irreducible representations of A. Conversely if x is in this intersection then $x \in (L : A)$ for all maximal left ideals L.
Hence $x = x1 \in L$ for all such L, so that $x \in \operatorname{Rad} A$.

(iii) It is obvious that $\operatorname{Sp}_{\pi(A)}\pi(x) \subset \operatorname{Sp}_A x$. Conversely let $\lambda \in \operatorname{Sp}_A x$. We
prove that there exists a continuous irreducible representation π of A on a Banach
space X such that $\lambda \in \operatorname{Sp}_{\mathcal{L}(X)}\pi(x) \subset \operatorname{Sp}_{\pi(A)}\pi(x)$. Replacing x by $\lambda 1 - x$, we may
suppose that x is not invertible in A. Suppose first that x has no left inverse in
A. Then there exists a maximal left ideal L containing x. Considering the left
regular representation associated to L, we have $\pi(x)\bar{1} = \bar{x} = \bar{0}$, so $\pi(x)$ is not
invertible in $\mathcal{L}(A/L)$. Suppose now that x has a left inverse ℓ in A. Of course
it does not have a right inverse. Let $u = 1 - x\ell$. Thus $ux = x - x(\ell x) = 0$. If
$u \in \operatorname{Rad} A$ then $x\ell = 1 - u$ is invertible, which implies that x has a right inverse
and hence a contradiction. Consequently $u \notin \operatorname{Rad} A$. By part (ii), there exists a
continuous irreducible representation π on a Banach space X such that $\pi(u) \neq 0$.
Since $\pi(u)\pi(x) = 0$ this implies that $\pi(x)$ is not invertible in $\mathcal{L}(X)$. $\square$

REMARK 1.   In fact the notions of bounded irreducible representations and continuous irreducible representations are the same (see [2], Theorem 7, p. 128).

REMARK 2.   If A is a commutative algebra then maximal left ideals coincide with maximal ideals and $(L : A) = L$. By Corollary 3.2.9, $A/L \simeq \mathbb{C}$. So in this case irreducible representations coincide with characters. Conversely, if all continuous irreducible representations of a Banach algebra are characters, then by Theorem 4.2.1 (ii), $A/\operatorname{Rad} A$ is commutative. So representation theory on Banach spaces with dimension greater than one is only interesting for non-commutative Banach algebras. In some sense Theorem 4.2.1 is a generalization of Theorem 4.1.2.

THEOREM 4.2.2 (I. SCHUR).   *Let A be a Banach algebra and let π be a continuous irreducible representation of A on a Banach space X. Then $C = \{T : T \in \mathcal{L}(X), T\pi(x) = \pi(x)T, \text{ for all } x \in A\}$ is isomorphic to $\mathbb{C}$.*

PROOF.   It is obvious that C is a closed subalgebra of $\mathcal{L}(X)$ containing the identity. Let $T \neq 0$ be in C. Then $T\pi(x)(X) = \pi(x)T(X) \subset T(X)$ for every $x \in A$, so $T(X)$ is invariant under π, and consequently $T(X) = X$. Also $\operatorname{Ker} T$ is invariant under π, but $\operatorname{Ker} T = X$ is impossible, so T is invertible as a linear operator. By the Open Mapping Theorem, T is invertible in $\mathcal{L}(X)$ and its inverse satisfies $T^{-1}\pi(x) = \pi(x)T^{-1}$ for all $x \in A$, so $T^{-1} \in C$. By Corollary 3.2.9, C is isomorphic to $\mathbb{C}$. $\square$

We now prove a very important theorem due to N. Jacobson. In fact this result can be extended to the general situation of non-commutative rings (see I. Herstein, *Noncommutative Rings*, 1968).

LEMMA 4.2.3.   *Let π be a continuous irreducible representation of A on a Banach space X. If ξ_1, ξ_2 are linearly independent in X there exists $a \in A$ such that $\pi(a)\xi_1 = 0$ and $\pi(a)\xi_2 \neq 0$.*

PROOF.   Suppose that $\pi(x)\xi_1 = 0$ implies $\pi(x)\xi_2 = 0$. Let $L_i = \{x : x \in A, \pi(x)\xi_i = 0\}$, for $i = 1, 2$. They are both maximal left ideals and $L_1 \subset L_2$, so $L_1 = L_2 = L$. The linear mappings T_i from A/L into X, defined for $i = 1, 2$ by $T_i(\bar{a}) = \pi(a)\xi_i$, are bounded and bijective. Let $D = T_2 T_1^{-1}$, which is a bounded linear operator on X. Let $\xi \in X$ and suppose that $\xi = \pi(b)\xi_1$. Then we have

$$\pi(a)D\xi = \pi(a)T_2 T_1^{-1}(\xi) = \pi(a)T_2(\bar{b}) = \pi(a)\pi(b)\xi_2 = \pi(ab)\xi_2,$$
$$D\pi(a)\xi = T_2 T_1^{-1}\pi(ab)\xi_1 = T_2(\overline{ab}) = \pi(ab)\xi_2,$$

so $D\pi(a) = \pi(a)D$ for all $a \in A$. By Theorem 4.2.2 there exists $\lambda \neq 0$ such that $D = \lambda I$. Then $T_2 = \lambda T_1$. So taking $a = 1$ we get $\xi_2 = \lambda \xi_1$ which gives a contradiction. $\square$

LEMMA 4.2.4. *Let π be a continuous irreducible representation of A on a Banach space X. If $\xi_1, \ldots, \xi_n$ are linearly independent in X there exists $a \in A$ such that $\pi(a)\xi_i = 0$ for $1 \leq i \leq n-1$ and $\pi(a)\xi_n \neq 0$.*

PROOF. This result is true for $n = 2$ by the previous lemma. We proceed by induction. Suppose $n > 2$ and that the property is true for $n-1$ linearly independent vectors in X. There exists $a_1 \in A$ such that $\pi(a_1)\xi_i = 0$ for $1 \leq i \leq n-2$ and $\pi(a_1)\xi_n \neq 0$. If $\pi(a_1)\xi_{n-1} = 0$, we have finished. If $\pi(a_1)\xi_{n-1}$ and $\pi(a_1)\xi_n$ are linearly independent then, by Lemma 4.2.3, there exists $a_2 \in A$ such that $\pi(a_2)\pi(a_1)\xi_{n-1} = 0$ and $\pi(a_2)\pi(a_1)\xi_n \neq 0$. So we take $a = a_2 a_1$. We now suppose that $\lambda \pi(a_1)\xi_{n-1} = \pi(a_1)\xi_n$ for some $\lambda \neq 0$. The vectors $\xi_1, \ldots, \xi_{n-2}, \lambda \xi_{n-1} - \xi_n$ are linearly independent so there exists $a_3 \in A$ such that $\pi(a_3)\xi_i = 0$ for $1 \leq i \leq n-2$ and $\pi(a_3)(\lambda \xi_{n-1} - \xi_n) \neq 0$. If $\pi(a_3)\xi_{n-1} = 0$, we have finished. So suppose $\pi(a_3)\xi_{n-1} \neq 0$. If $\pi(a_3)\xi_{n-1}$ and $\pi(a_3)\xi_n$ are linearly independent there exists $a_4 \in A$ such that $\pi(a_4)\pi(a_3)\xi_{n-1} = 0$ and $\pi(a_4)\pi(a_3)\xi_n \neq 0$. Then we take $a = a_4 a_3$. So suppose once again that we have $\mu\pi(a_3)\xi_{n-1} = \pi(a_3)\xi_n$. By hypothesis, $\lambda \pi(a_3)\xi_{n-1} \neq \pi(a_3)\xi_n$, so $\lambda \neq \mu$. Because $\pi(a_3)\xi_{n-1} \neq 0$ there exists $a_5 \in A$ such that $\pi(a_5)\pi(a_3)\xi_{n-1} = \pi(a_1)\xi_{n-1}$. Taking $a = a_1 - a_5 a_3$ we have $\pi(a)\xi_i = 0$ for $1 \leq i \leq n-1$ and $\pi(a)\xi_n = \pi(a_1)\xi_n - \pi(a_5)\pi(a_3)\xi_n = \lambda\pi(a_1)\xi_{n-1} - \mu\pi(a_5 a_3)\xi_{n-1} = (\lambda - \mu)\pi(a_1)\xi_{n-1} \neq 0$. $\square$

THEOREM 4.2.5 (JACOBSON DENSITY THEOREM). *Let π be a continuous irreducible representation of A on a Banach space X. If $\xi_1, \ldots, \xi_n$ are linearly independent in X and if $\eta_1, \ldots, \eta_n$ are in X there exists $a \in A$ such that $\pi(a)\xi_i = \eta_i$, for $i = 1, \ldots, n$.*

PROOF. By Lemma 4.2.4 there exists $b_k \in A$ such that $\pi(b_k)\xi_i = 0$ if $i \neq k$ and $\pi(b_k)\xi_k \neq 0$. Then there exists $c_k \in A$ such that $\pi(c_k)\pi(b_k)\xi_k = \eta_k$. We take $a = c_1 b_1 + \cdots + c_n b_n$. $\square$

If π is an irreducible representation of A over a vector space X of dimension n, then $\pi(A)$ is isomorphic to $M_n(\mathbf{C})$ (see Exercise IV.13).

The next corollary is very useful.

COROLLARY 4.2.6 (A. SINCLAIR). *With the hypotheses of Theorem 4.2.5, we suppose further that $\eta_1, \ldots, \eta_n$ are linearly independent. Then there exists a invertible in A such that $\pi(a)\xi_i = \eta_i$ for $i = 1, \ldots, n$.*

PROOF. Let F be the finite-dimensional linear subspace of A generated by $\xi_1, \ldots, \xi_n, \eta_1, \ldots, \eta_n$. Given the hypothesis, there exists a linear mapping T from F onto F, which is invertible in $\mathfrak{L}(F)$, such that $T\xi_i = \eta_i$, for $i = 1, \ldots, n$. The algebra $\mathfrak{L}(F)$ is isomorphic to a matrix algebra $M_k(\mathbb{C})$ for some $k \leq 2n$, so it is a Banach algebra. Because $\mathrm{Sp}_{\mathfrak{L}(F)}T$ is finite, by Theorem 3.3.6, there exists $R \in \mathfrak{L}(F)$ such that $T = e^R$. Let B be a basis of F containing $\xi_1, \ldots, \xi_n$. By theorem 4.2.5, there exists $a \in A$ such that $\pi(a)\xi = R\xi$ for all $\xi \in B$. Consequently $\pi(a)$ and R coincide on F. In particular, $\pi(a)^k$ and R^k coincide on F for all integer $k \geq 1$. So, by continuity of π, we get

$$\pi(e^a)\xi_i = e^R\xi_i = T\xi_i = \eta_i \ , \quad \text{for } i = 1, \ldots, n$$

and e^a is invertible in A. $\square$

To finish this section we give a result which was proved by I. Kaplansky in his book *Infinite Abelian Groups*, Ann Arbor, 1969, using the theory of finitely generated modules.

Let T be a linear operator on a complex vector space X. We say that T is *locally algebraic* if for every $\xi \in X$ there exists a non-trivial polynomial p such that $p(T)\xi = 0$. The standard result of I. Kaplansky states that *boundedly locally algebraic* (the degree of p is bounded independently of ξ) implies *algebraic* (for another proof see the book by H. Radjavi and P. Rosenthal, *Invariant Subspaces*, Berlin, 1973). This important result has many consequences. We present an analytic proof of the result which is very interesting because it implies a surprising generalization.

THEOREM 4.2.7 (I. KAPLANSKY). *Let X be a complex vector space and let T be a linear operator from X into X. Suppose that there exists an integer $n \geq 1$ such that $\xi, T\xi, \ldots, T^n\xi$ are linearly dependent for all $\xi \in X$. Then T is algebraic of degree less than or equal to n.*

PROOF. Suppose that n is the smallest integer having this property. Hence there exists $\xi_0 \in X$ such that $\xi_0, T\xi_0, \ldots, T^{n-1}\xi_0$ are linearly independent but $\xi_0, T\xi_0, \ldots, T^n\xi_0$ are not. Then there exists a monic polynomial p_0 of degree n such that $p_0(T)\xi_0 = 0$ and if p is another monic polynomial of degree n

such that $p(T)\xi_0 = 0$, then $p = p_0$. Let $\eta \in X$ be an arbitrary fixed vector. We now prove that $p_0(T)\eta = 0$. Let F be the linear subspace generated by $\xi_0, T\xi_0, \ldots, T^n\xi_0, \eta, T\eta, \ldots, T^n\eta$. Then $\dim F \leq 2n$. For $\lambda \in \mathbf{C}$ we set

$$
\left\{
\begin{aligned}
f_0(\lambda) &= \xi_0 + \lambda\eta \in F \\
f_1(\lambda) &= Tf_0(\lambda) \in F \\
&\;\;\vdots \\
f_{n-1}(\lambda) &= T^{n-1}f_0(\lambda) \in F \\
g(\lambda) &= T^n f_0(\lambda) \in F.
\end{aligned}
\right.
$$

Because $f_0(0), \ldots, f_{n-1}(0)$ are linearly independent in F there exist n linear functionals on F, denoted by $\phi_0, \ldots, \phi_{n-1}$, such that

$$
\phi_i(f_j(0)) = \delta_{ij} , \quad \text{for } 0 \leq i, j \leq n - 1. \tag{1}
$$

We define

$$
\Delta(\lambda) =
\begin{vmatrix}
\phi_0(f_0(\lambda)) & \phi_0(f_1(\lambda)) & \cdots & \phi_0(f_{n-1}(\lambda)) \\
\phi_1(f_0(\lambda)) & \phi_1(f_1(\lambda)) & \cdots & \phi_1(f_{n-1}(\lambda)) \\
\vdots & \vdots & \ddots & \vdots \\
\phi_{n-1}(f_0(\lambda)) & \phi_{n-1}(f_1(\lambda)) & \cdots & \phi_{n-1}(f_{n-1}(\lambda))
\end{vmatrix}
$$

which is a polynomial of degree $\leq n$, satisfying $\Delta(0) = 1$. Let E be the finite set of its zeros. From the hypothesis we conclude that for $\lambda \notin E$ there exist $\alpha_0(\lambda), \ldots, \alpha_{n-1}(\lambda) \in \mathbf{C}$ such that

$$
g(\lambda) = \alpha_0(\lambda)f_0(\lambda) + \cdots + \alpha_{n-1}(\lambda)f_{n-1}(\lambda), \tag{2}
$$

so we have

$$
\left\{
\begin{aligned}
\phi_0(g(\lambda)) &= \alpha_0(\lambda)\phi_0(f_0(\lambda)) + \cdots + \alpha_{n-1}(\lambda)\phi_0(f_{n-1}(\lambda)) \\
&\;\;\vdots \\
\phi_{n-1}(g(\lambda)) &= \alpha_0(\lambda)\phi_{n-1}(f_0(\lambda)) + \cdots + \alpha_{n-1}(\lambda)\phi_{n-1}(f_{n-1}(\lambda)).
\end{aligned}
\right. \tag{3}
$$

By Cramer's formulae the α_i coincide on $\mathbf{C}\backslash E$ with rational functions. Relation (2) can be written:

$$
\left\{
\begin{aligned}
&p_\lambda(T)f_0(\lambda) = 0 , \quad \text{for } \lambda \notin E \text{ with} \\
&p_\lambda(T) = T^n - \alpha_{n-1}(\lambda)T^{n-1} - \cdots - \alpha_0(\lambda)\mathbf{1}.
\end{aligned}
\right. \tag{4}
$$

Let us denote by $\beta_1(\lambda), \cdots, \beta_n(\lambda)$ the roots of the polynomial p_λ. We have

$$(T - \beta_1(\lambda)\mathbf{1}) \cdots (T - \beta_n(\lambda)\mathbf{1})f_0(\lambda) = 0 , \quad \text{for } \lambda \notin E, \tag{5}$$

and obviously $(T - \beta_2(\lambda)\mathbf{1}) \cdots (T - \beta_n(\lambda)\mathbf{1}f_0(\lambda) \neq 0$ for $\lambda \notin E$, by the definition of E. So (5) implies that $\beta_1(\lambda)$ is in the spectrum of T. A similar argument implies that $\beta_2(\lambda), \ldots, \beta_n(\lambda)$ are also in the spectrum of T. Consequently $|\beta_i(\lambda)| \leq \|T\|$, for $i = 1, \ldots, n$ and $\lambda \notin E$, where $\|T\|$ is the operator norm corresponding to a given norm on the invariant subspace F. So the symmetric functions $\alpha_0(\lambda), \ldots, \alpha_{n-1}(\lambda)$ are also bounded on $\mathbf{C} \backslash E$. Because the α_i coincide with rational functions on $\mathbf{C} \backslash E$, we conclude from Liouville's theorem that there are constant numbers $\gamma_0, \ldots, \gamma_{n-1} \in \mathbf{C}$ such that $\alpha_i(\lambda) = \gamma_i$, for $\lambda \notin E$. Let $p(z) = z^n - \gamma_{n-1}z^{n-1} - \cdots - \gamma_0$. Then $p(T)f_0(\lambda) = 0$ on $\mathbf{C} \backslash E$, but also on $\mathbf{C}$, by continuity in λ. In particular $p(T)\xi_0 = 0$, so $p = p_0$. Consequently $p_0(T)\eta = 0$, for an arbitrary $\eta \in X$. Hence $p_0(T) = 0$, so T is algebraic of degree $\leq n$. $\square$

COROLLARY 4.2.8. *Let X be a complex Banach space and let T be a bounded linear operator from X into X. Suppose that for every ξ in X there exists an integer n (depending on ξ) such that $\xi, T\xi, \ldots, T^n\xi$ are linearly dependent. Then T is algebraic.*

PROOF.   For $k \geq 1$ let $X_k = \{\xi : \xi \in X$ and $\xi, T\xi, \ldots, T^k\xi$ are linearly dependent$\}$. Then X_k is the set of ξ for which there exists a monic polynomial p with degree $\leq k$, such that $p(T)\xi = 0$. By continuity of T and Lemma 2.2.1, the sets X_k are closed. By hypothesis $X = \cup_{k=1}^{\infty} X_k$. Consequently, by Theorem 1.1.1, there exist a ball $B(\xi_0, r)$ and a smallest integer m such that $\xi \in X_m$, for $\|\xi - \xi_0\| < r$. Let $\eta \in X$ be a fixed arbitrary vector. Then $\xi_0 + t\eta \in B(\xi_0, r)$ if $|t| \cdot \|\eta\| < r$. With this hypothesis there exists a monic polynomial p of degree $\leq m$ such that $p(T)(\xi_0 + t\eta) = 0$. Moreover there exists a monic polynomial p_0 of degree $\leq m$ such that $p_0(T)\xi_0 = 0$. Hence $p_0(T)p(T)\eta = 0$. So $\eta \in X_{2m}$. We then apply Theorem 4.2.7 with $n = 2m$ to get the result. $\square$

The proof of the next theorem is a slight modification of the argument used in the proof of Theorem 4.2.7.

THEOREM 4.2.9. *Let X and Y be two complex vector spaces and let $T_1, \ldots, T_n$ be linear operators from X into Y. Suppose that for every $\xi \in X$ the vectors $T_1\xi, \ldots, T_n\xi$ are linearly dependent. Then there exist $\alpha_1, \ldots, \alpha_n \in \mathbf{C}$ not all zero such that $S = \alpha_1 T_1 + \cdots + \alpha_n T_n$ has finite rank $\leq n - 1$. Moreover, if $X = Y$ and the T_i commute, then $S^2 = 0$.*

PROOF. If for all $\xi \in X$, the vectors $T_1\xi, \ldots, T_{n-1}\xi$ are linearly dependent, it is enough to prove the result with $T_1, \ldots, T_{n-1}$. So suppose that there exists $\xi_0 \in X$ such that $T_1\xi_0, \ldots, T_{n-1}\xi_0$ are linearly independent and $T_1\xi_0, \ldots, T_n\xi_0$ are not. Then there exist $\alpha_1, \ldots, \alpha_{n-1} \in \mathbb{C}$ such that

$$(T_n + \alpha_{n-1}T_{n-1} + \cdots \alpha_1 T_1)\xi_0 = 0. \tag{6}$$

Let $\eta \in X$ be an arbitrary fixed vector and let F be the linear subspace of Y generated by $T_1\xi_0, \ldots, T_n\xi_0, T_1\eta, \ldots, T_n\eta$. Then $\dim F \leq 2(n-1)$. For $\lambda \in \mathbb{C}$ we set

$$\begin{cases} f_0(\lambda) = \xi_0 + \lambda\eta \\ f_1(\lambda) = T_1 f_0(\lambda) \in F \\ \qquad \vdots \\ f_{n-1}(\lambda) = T_{n-1} f_0(\lambda) \in F \\ g(\lambda) = T_n f_0(\lambda) \in F. \end{cases} \tag{7}$$

Because $f_1(0), \ldots, f_{n-1}(0)$ are linearly independent in F there exist $n-1$ linear functionals on F, denoted by $\phi_1, \ldots, \phi_{n-1}$, such that

$$\phi_i(f_j(0)) = \delta_{ij}, \quad \text{for } 1 \leq i, j \leq n-1. \tag{8}$$

We define

$$\Delta(\lambda) = \begin{vmatrix} \phi_1(f_1(\lambda)) & \phi_1(f_2(\lambda)) & \cdots & \phi_1(f_{n-1}(\lambda)) \\ \phi_2(f_1(\lambda)) & \phi_2(f_2(\lambda)) & \cdots & \phi_2(f_{n-1}(\lambda)) \\ \vdots & \vdots & \ddots & \vdots \\ \phi_{n-1}(f_1(\lambda)) & \phi_{n-1}(f_2(\lambda)) & \cdots & \phi_{n-1}(f_{n-1}(\lambda)) \end{vmatrix}$$

which is a polynomial of degree $\leq n-1$, satisfying $\Delta(0) = 1$, and

$$\Delta_i(\lambda) = \begin{vmatrix} \phi_1(f_1(\lambda)) & \cdots & \phi_1(g(\lambda)) & \cdots & \phi_1(f_{n-1}(\lambda)) \\ \phi_2(f_1(\lambda)) & \cdots & \phi_2(g(\lambda)) & \cdots & \phi_2(f_{n-1}(\lambda)) \\ \vdots & \ddots & \vdots & \ddots & \vdots \\ \phi_{n-1}(f_1(\lambda)) & \cdots & \phi_{n-1}(g(\lambda)) & \cdots & \phi_{n-1}(f_{n-1}(\lambda)) \end{vmatrix}$$
$$\underset{i^{\text{th}}\text{column}}{\uparrow}$$

which is also a polynomial of degree $\leq n-1$, satisfying $-\Delta_i(0) = \alpha_i$, by (6) and (8). If E denotes the set of zeros of Δ then, arguing as in the proof of Theorem 4.2.7, we conclude that

$$(\Delta(\lambda)T_n - \Delta_{n-1}(\lambda)T_{n-1} - \cdots - \Delta_1(\lambda)T_1)f_0(\lambda) = 0 \tag{9}$$

on $\mathbf{C} \backslash E$, but, by continuity in λ, this relation is true on all $\mathbf{C}$. Let $\alpha_n = 1$ and let $\beta_1, \ldots, \beta_n$ be the coefficients of λ respectively in $-\Delta_1(\lambda), \ldots, -\Delta_{n-1}(\lambda), \Delta(\lambda)$. Setting $S = \alpha_1 T_1 + \cdots + \alpha_n T_n$ (which does not depend on η!), $R = \beta_1 T_1 + \cdots + \beta_n T_n$ (which depends on η!) and looking at the coefficients of degree 0 and 1 in λ, from (9) we obtain:

$$\begin{cases} S\xi_0 = 0 \\ S\eta + R\xi_0 = 0. \end{cases}$$

Consequently $S\eta$ is in the linear subspace generated by $T_1\xi_0, \ldots, T_{n-1}\xi_0$. So S has finite rank $\leq n - 1$. If moreover the T_i commute, then S and R commute, so $S^2\eta = -SR\xi_0 = -RS\xi_0 = 0$. Hence $S^2 = 0$. $\square$

REMARK 3.   Let P and Q be two different projections, having the same range of dimension 1, defined on a complex vector space X. For every $\xi \in X$ the vectors $P\xi$ and $Q\xi$ are dependent, and obviously there are linear combinations of P and Q having rank one. But $\alpha P + \beta Q \neq 0$ for every $\alpha, \beta \in \mathbf{C}$. So in general it is impossible to have $S = 0$ in Theorem 4.2.9.

*EXERCISE 1.   Prove that $M_n(\mathbb{C})$ and $\mathcal{L}(H)$, where H is a Hilbert space, have no characters.

EXERCISE 2.   Let A be a commutative Banach algebra and let $x \in A$. Prove that there exists a character χ such that $\rho(x) = |\chi(x)|$.

EXERCISE 3.   Prove Corollary 3.2.10 and Theorem 3.4.1 using Theorem 4.1.2.

EXERCISE 4.   Let K_1, K_2 be two compact spaces. Suppose that there exists a linear isometry from $C(K_1)$ onto $C(K_2)$. Prove that K_1 and K_2 are homeomorphic.

EXERCISE 5.   Let K be a compact subset of $\mathbb{C}^n$. Prove that $\mathfrak{M}(P(K))$ is homeomorphic to $\hat{K}$. Suppose now that every continuous function on K can be uniformly approximated on K by polynomials. Conclude that $K = \hat{K}$.

***EXERCISE 6.   Give an example of a curve Γ in $\mathbb{C}^n$ $(n \geq 2)$ such that $\hat{\Gamma} \neq \Gamma$. (If you do not succeed look at [10]). For such an example conclude that $C(\Gamma) \neq P(\Gamma)$.

EXERCISE 7.   Let S be a completely regular Hausdorff space and let $C_b(S)$ be the algebra of bounded continuous complex-valued functions on S. Prove that $C_b(S)$ is a Banach algebra for the norm $\|f\| = \sup_{x \in S} |f(x)|$. Prove that $C_b(S)$ is isomorphic to $C(X)$, for some compact set X such that S is homeomorphic to a dense subset of X, and every bounded continuous complex-valued function on S extends continuously to X (X is called the *Stone-Čech compactification* of S).

EXERCISE 8.   Let K be a compact and convex subset of the complex plane. Prove that K is the intersection of all closed disks containing K.

EXERCISE 9.   Let D_1 and D_2 be two domains of the complex plane. Characterize explicitly all the linear isometries from $A(D_1)$ onto $A(D_2)$.

EXERCISE 10.   Let A be a commutative Banach algebra. A *boundary* for A is a closed subset E of $\mathfrak{M}(A)$ such that $\sup_{\chi \in E} |\chi(f)| = \sup_{\chi \in \mathfrak{M}(A)} |\chi(f)|$, for every $f \in A$. Prove that the intersection of all boundaries of A is a boundary of A (called the *Šilov boundary* of A). Given a compact subset K of $\mathbb{C}$ determine explicitly the Šilov boundaries of $C(K)$ and $A(K)$.

**EXERCISE 11.   Let K be a compact set and let $|\cdot|$ be a submultiplicative norm on $C(K)$ (which is not complete). Prove that $\|f\|_\infty \leq |f|$, for all $f \in C(K)$. This important result is due to I. Kaplansky. If you do not succeed look at the book by A. Sinclair mentioned at the end of Chapter V, §5.

***EXERCISE 12.   Try to prove the following result of H. Rossi. Let A be a commutative Banach algebra and fix $\chi_0 \in \mathfrak{M}(A)\backslash S$, where S denotes the Šilov boundary of A. Let U be a neighbourhood of χ_0 with $U \subset \mathfrak{M}(A)\backslash S$. Then for all $f \in A$ we have

$$|\chi_0(f)| \le \sup_{\chi \in \partial U} |\chi(f)|.$$

If you do not succeed please read [10], Chapter 9.

EXERCISE 13.   Let A be a Banach algebra and let π be an irreducible representation of A on a linear vector space of dimension n. Prove that $\pi(A)$ is isomorphic to $M_n(\mathbf{C})$.

*EXERCISE 14.   Is it possible to extend Theorem 4.2.9 to a countable family of linear operators from X into Y?

Chapter V

SOME APPLICATIONS
OF SUBHARMONICITY

The powerful technique of subharmonic functions which we introduced in Chapter III, §4, has a great number of applications in spectral theory.

§1. Some Elementary Applications

We start with a simple application which improves Theorem 3.4.2 and the Geršgorin theorem for matrices.

THEOREM 5.1.1. *Let a be an element of a Banach algebra and let U be a bounded open set containing $\mathrm{Sp}\,a$. Then $\sup_{\lambda \in \partial U} \rho((a-\lambda 1)^{-1}(x-a)) < 1$ implies $\mathrm{Sp}\,x \subset U$.*

PROOF. If $\lambda \in \mathbf{C} \backslash \overline{U}$, then $a - \lambda 1$ is invertible, so we define $f(\lambda) = (a-\lambda 1)^{-1}(x-a)$. This function is analytic on $\mathbf{C} \backslash \overline{U}$ and goes to 0 at infinity. By the Maximum Principle for subharmonic functions on unbounded open sets (see the remark following Corollary A.1.4), we conclude from the hypothesis that $\rho(f(\lambda)) < 1$ on $\mathbf{C} \backslash \overline{U}$. So $1 + f(\lambda)$ is invertible for $\lambda \notin U$. Hence $(a - \lambda 1)(1 + f(\lambda)) = x - \lambda 1$ is invertible for $\lambda \notin U$. Consequently $\mathrm{Sp}\,x \subset U$. $\square$

COROLLARY 5.1.2 (S.A.GERŠGORIN). *Let $x = (a_{ij})$ be a $n \times n$ matrix and let a be the diagonal matrix with $a_{11}, \ldots, a_{nn}$ on the diagonal. Suppose that for $\epsilon > 0$, the disks $B(a_{11}, \epsilon), \ldots, B(a_{nn}, \epsilon)$ have disjoint or identical boundaries (this last case occurs if $a_{ii} = a_{jj}$ for $i \neq j$). Then $\|x - a\| < \epsilon$ implies $\mathrm{Sp}\,x \subset B(a_{11}, \epsilon) \cup \cdots \cup B(a_{nn}, \epsilon)$.*

PROOF. We have $\rho((a - \lambda 1)^{-1}(x - a)) \leq \|(a - \lambda 1)^{-1}\| \cdot \|x - a\|$. But the diagonal matrix $(a - \lambda 1)^{-1}$ has the diagonal coefficients $1/(a_{11} - \lambda), \ldots, 1/(a_{nn} - \lambda)$, and hence $\|(a - \lambda 1)^{-1}\| = \max_i 1/|a_{ii} - \lambda| = 1/\epsilon$ when λ lies on the boundary of $U = B(a_{11}, \epsilon) \cup \cdots \cup B(a_{nn}, \epsilon)$. We then apply Theorem 5.1.1. $\square$

THEOREM 5.1.3 (D.C.KLEINECKE-F.V.ŠIROKOV). *Let a, b be in a Banach algebra. Suppose that $a(ab - ba) = (ab - ba)a$. Then $\rho(ab - ba) = 0$.*

PROOF. Let $[x, y]$ be the commutator $xy - yx$ of x and y. Then $e^{\lambda a} b e^{-\lambda a} = b + \lambda[a, b] + \frac{\lambda^2}{2!}[a, [a, b]] + \cdots = b + \lambda[a, b]$, for all $\lambda \in \mathbf{C}$. But $\mu \mapsto \rho(\mu b + [a, b])$ is subharmonic on $\mathbf{C}$, so by Theorem A.1.2 we have

$$\rho([a, b]) = \limsup_{\substack{\mu \to 0 \\ \mu \neq 0}} \rho(\mu b + [a, b]) = 0. \ \square$$

THEOREM 5.1.4. *Let a, b be in a Banach algebra. Suppose that $(ab - ba)a = 0$, or that $a(ab - ba) = 0$, and that 0 is in the boundary of some component of $\mathbf{C} \setminus \mathrm{Sp}\, a$. Then $\rho(ab - ba) = 0$.*

PROOF. Suppose for instance that $(ab - ba)a = 0$, the other case being studied similarly. Then we have $(\lambda 1 - a)b = (b - \frac{1}{\lambda}(ab - ba))(\lambda 1 - a)$ for all $\lambda \neq 0$. Hence $(\lambda 1 - a)b(\lambda 1 - a)^{-1} = b - \frac{1}{\lambda}(ab - ba)$, for all $\lambda \notin \mathrm{Sp}\, a$. Let U be the component of $\mathbf{C} \setminus \mathrm{Sp}\, a$ such that $0 \in \partial U$. Then

$$|\lambda| \rho(b) = \rho(\lambda b - (ab - ba)) \quad \text{for } \lambda \in U.$$

Because $\lambda \mapsto \rho(\lambda b - (ab - ba))$ is subharmonic, we conclude from Theorem A.1.2 that

$$\rho(ab - ba) = \limsup_{\substack{\lambda \in U \\ \lambda \to 0}} \rho(\lambda b - (ab - ba)) = 0. \ \square$$

In particular, if T is a compact operator on an infinite-dimensional Banach space, then $(TS - ST)T = 0$ implies $\rho(TS - ST) = 0$.

§2. Spectral Characterizations of Commutative Banach Algebras

If A is commutative we know from Corollary 3.2.10 and Theorem 3.4.1 that the spectral radius is subadditive, submultiplicative and uniformly continuous on A. By Theorem 3.1.5, the same result is also true supposing $A/\mathrm{Rad}\,A$ to be commutative. Surprisingly, the converse is true.

First we strengthen a lemma due to C. Le Page.

Given a Banach algebra A, then by definition $Z(A)$, the *centre modulo the radical* of A, is the set of $a \in A$ such that $ax - xa \in \mathrm{Rad}\,A$ for all $x \in A$.

THEOREM 5.2.1.   *Let $a \in A$ be such that $\# \operatorname{Sp}(ax - xa) = 1$ for all $x \in A$. Then $a \in Z(A)$.*

PROOF.   Let π be a continuous irreducible representation of A on a Banach space X. Suppose there exists $\xi \in X$ such that ξ, $\eta = \pi(a)\xi$ and $\pi(a)\eta$ are linearly independent. Then, by Theorem 4.2.5, there exists $x \in A$ such that

$$\begin{cases} \pi(x)\xi = 0 \\ \pi(x)\eta = -\xi \\ \pi(x)\pi(a)\eta = -\eta. \end{cases}$$

Then $\pi(ax - xa)\xi = -\pi(x)\eta = \xi$ and $\pi(ax - xa)\eta = -\eta + \eta = 0$. So we have $\{0,1\} \subset \operatorname{Sp}\pi(ax - xa) \subset \operatorname{Sp}(ax - xa)$ which is a contradiction. Consequently for $\xi \in X$, the vectors ξ, $\pi(a)\xi$ and $\pi(a)^2\xi$ are linearly dependent. By Theorem 4.2.7, $\pi(a)$ is algebraic of degree ≤ 2. Without loss of generality we may suppose that $\pi(a)^2 = \gamma 1$, for some $\gamma \in \mathbb{C}$. As a matter of fact if we have $\pi(a)^2 = \alpha\pi(a) + \beta 1$, then taking $a' = a - \frac{\alpha}{2}1$ we have $\operatorname{Sp}(a'x - xa') = \operatorname{Sp}(ax - xa)$, for all $x \in A$, and $\pi(a')^2 = (\beta + \alpha^2/4)1$. If $\pi(a)$ is not algebraic of degree 1, there exists $\xi \in X$ such that ξ and $\eta = \pi(a)\xi$ are linearly independent. By Theorem 4.2.5 there exists $x \in A$ such that

$$\begin{cases} \pi(x)\xi = \xi \\ \pi(x)\eta = \xi + \eta. \end{cases}$$

Then $\pi(ax - xa)\xi = \eta - (\xi + \eta) = -\xi$ and $\pi(ax - xa)\eta = \pi(a)(\xi + \eta) - \pi(x)\gamma\xi = \eta + \gamma\xi - \gamma\xi = \eta$. Consequently $\{-1,1\} \subset \operatorname{Sp}\pi(ax - xa) \subset \operatorname{Sp}(ax - xa)$ which gives a contradiction. Hence $\pi(a)$ is algebraic of degree 1, that is $\pi(a) = \alpha 1$ for some $\alpha \in \mathbb{C}$. Then $\pi(ax - xa) = 0$ for all continuous irreducible representations π. So by Theorem 4.2.1 (ii), we have $ax - xa \in \operatorname{Rad} A$, for all $x \in A$. $\square$

THEOREM 5.2.2.   *Let $a \in A$. Then the following properties are equivalent:*

(i) *$a \in Z(A)$,*

(ii) *there exists $M > 0$ such that $\rho(a + x) \leq M(1 + \rho(x))$, for every $x \in A$,*

(iii) *there exists $N > 0$ such that $\rho((a - \lambda 1)^{-1}x) \leq N\rho((a - \lambda 1)^{-1})\rho(x)$ for every $x \in A$ and $\lambda \notin \operatorname{Sp} a$.*

PROOF.   Changing A for $A/\operatorname{Rad} A$ if necessary, we may suppose without loss of generality that A is semi-simple. Then (i) implies (ii) and (iii), by Corollary 3.2.10, taking $M \geq \max(1, \rho(a))$ and $N \geq 1$. We now prove that (ii) implies (i). Let $u \in A$ be fixed and $\lambda \in \mathbb{C}$. We define an analytic function from $\mathbb{C}$ into A by

$$f(\lambda) = \begin{cases} \dfrac{a - e^{\lambda u} a e^{-\lambda u}}{\lambda} &, \text{ for } \lambda \neq 0 \\ [a, u] &, \text{ for } \lambda = 0. \end{cases}$$

For $\lambda \neq 0$ we have $\rho(f(\lambda)) \leq \frac{M}{|\lambda|}(1 + \rho(e^{\lambda u} a e^{-\lambda u})) = \frac{M}{|\lambda|}(1 + \rho(a))$. So the sub-harmonic function $\lambda \mapsto \rho(f(\lambda))$ goes to zero at infinity. Hence it is identically 0. Consequently $\rho(au - ua) = 0$, for all $u \in A$. By Theorem 5.2.1, $a \in Z(A)$. If we look at the proof of Theorem 3.4.1 we conclude that (iii) implies $\operatorname{Sp} y \subset \operatorname{Sp} a + N\rho(y - a)$, for every $y \in A$. In particular we have $\rho(a + x) \leq \rho(a) + N\rho(x) \leq M(1 + \rho(x))$ for $M \geq \max(\rho(a), N)$. So (iii) implies (ii). $\square$

COROLLARY 5.2.3.   *Let A be a Banach algebra. Then the following properties are equivalent:*

(i) *$A/\operatorname{Rad} A$ is commutative,*

(ii) *ρ is subadditive on A, that is there exists $M > 0$ such that $\rho(x + y) \leq M(\rho(x) + \rho(y))$, for all $x, y \in A$,*

(iii) *ρ is submultiplicative on A, that is there exists $N > 0$ such that $\rho(xy) \leq N\rho(x)\rho(y)$, for all $x, y \in A$,*

(iv) *ρ is uniformly continuous on A, which implies that there exists $C > 0$ such that $|\rho(x) - \rho(y)| \leq C\|x - y\|$, for all $x, y \in A$.*

We now give another characterization of $Z(A)$ due to S. Grabiner, along with a new proof. We recall that δ denotes the diameter.

THEOREM 5.2.4.   *We have $a \in Z(A)$ if and only if $\sup_{x \in A} \delta(e^x a e^{-x} - a) < +\infty$.*

PROOF.   Suppose that a and b commute. Then we have $\delta(a + b) \leq 2\rho(a + b) \leq 2(\rho(a) + \rho(b))$ by Corollary 3.2.10. In fact it is possible to prove that $\delta(a + b) \leq \delta(a) + \delta(b)$, see Exercise III.11. Suppose $a \in Z(A)$. Then $\delta(e^x a e^{-x} - a) \leq 2(\rho(a) + \rho(e^x a e^{-x})) = 4\rho(a)$, using the remark just before Theorem 3.2.4. We now prove the converse. Let $u \in A$ be fixed and let $\lambda \in \mathbb{C}$. Then $\lambda \mapsto \delta(f(\lambda))$ is subharmonic on $\mathbb{C}$ by Theorem 3.4.24, $f(\lambda)$ being defined as in the proof of Theorem 5.2.2. But it goes to zero at infinity. So it is identically zero. Consequently $\delta([a, u]) = 0$, or equivalently $\# \operatorname{Sp}(au - ua) = 1$, for all $u \in A$. Hence we have $a \in Z(A)$ by Theorem 5.2.1. $\square$

§3. Spectral Characterizations of the Radical

THEOREM 5.3.1 (J.ZEMÁNEK).   *Let A be a Banach algebra. Then the following properties are equivalent:*

(i) *a is in the Jacobson radical of A,*

(ii) *$\operatorname{Sp}(a + x) = \operatorname{Sp} x$, for all $x \in A$,*

(iii) *$\rho(a + x) = 0$, for all quasi-nilpotent elements x in A,*

(iv) *$\rho(a + x) = 0$, for all quasi-nilpotent elements x in a neighbourhood of 0 in A,*

(v) *there exists $C > 0$ such that $\rho(x) \leq C\|x - a\|$, for all x in a neighbourhood of a in A.*

PROOF.   It is easy to see that (i) implies (ii) and (v) (with $C = 1$). It is obvious that (ii) implies (iii), which implies (iv). We now prove that (iii) implies (i). Taking $x = 0$ we have $\rho(a) = 0$, and so $\rho(e^x a e^{-x}) = 0$, for all $x \in A$. Consequently $\rho(a - e^x a e^{-x}) = 0$, for all $x \in A$. By Theorem 5.2.4, we have $a \in Z(A)$. For every continuous irreducible representation π we have $\pi(a) = \alpha 1$, for some $\alpha \in \mathbf{C}$, and moreover $\rho(\pi(a)) = 0$. Hence $\pi(a) = 0$. By Theorem 4.2.1 (ii), $a \in \operatorname{Rad} A$. We now prove that (iv) implies (iii). Suppose that $\rho(a + x) = 0$ for all quasi-nilpotent elements x such that $\|x\| < r$. Let $q \neq 0$ be an arbitrary quasi-nilpotent element. Then $\rho(a + \lambda q) = 0$ for $|\lambda| \cdot \|q\| < r$. But $\lambda \mapsto \log \rho(a + \lambda q)$ is subharmonic and is $-\infty$ on the disk $\{\lambda : |\lambda| < r/\|q\|\}$, so by H. Cartan's theorem it is $-\infty$ everywhere. Hence $\rho(a + q) = 0$. To finish we prove that (v) implies (iv). Let q be a quasi-nilpotent element of A. We have $\rho(\lambda a + q) = |\lambda|\rho(a + q/\lambda) \leq C\|q\|$, for $|\lambda|$ large enough. So the subharmonic function $\lambda \mapsto \rho(\lambda a + q)$ is bounded on $\mathbf{C}$, and hence constant. Consequently $\rho(a + q) = \rho(q) = 0$. $\square$

Property (v) implies that if a quasi-nilpotent element a is not in the radical of A then the spectral radius is continuous but not lipschitzian at a.

THEOREM 5.3.2.   *Let A be a semi-simple Banach algebra and let $a \in A$. There exists $\alpha \in \mathbf{C}$ such that $a = \alpha 1$ if and only if $\# \operatorname{Sp}(a + q) = 1$ for all quasi-nilpotent elements q of A.*

PROOF.   If $a = \alpha 1$ then $\operatorname{Sp}(a + q) = \{\alpha\}$ for all q quasi-nilpotent. We prove the converse. Let q be a fixed quasi-nilpotent element. For $\lambda \in \mathbf{C}$ there exists $h(\lambda) \in \mathbf{C}$ such that $\operatorname{Sp}(a + \lambda q) = \{h(\lambda)\}$. By Corollary 3.4.18, h is entire. But $\limsup_{|\lambda| \to \infty} \left| \frac{h(\lambda)}{\lambda} \right| \leq \rho(q) = 0$ because $\frac{h(\lambda)}{\lambda} \in \operatorname{Sp}(\frac{a}{\lambda} + q)$ for $\lambda \neq 0$. So, by Liouville's theorem for entire functions, $h(\lambda)$ is constant. Consequently $\mathrm{Sp}(a + \lambda q) = \mathrm{Sp}\, a$, for $\lambda \in \mathbf{C}$. But $\mathrm{Sp}\, a = \{\alpha\}$ implies $a = \alpha 1 + q_0$ for some quasi-nilpotent element q_0. So $\mathrm{Sp}(q_0 + q) = \{0\}$ for all quasi-nilpotent elements q. By Theorem 5.3.1 (iii), $q_0 \in \mathrm{Rad}\, A = \{0\}$ and so $a = \alpha 1$. $\square$

Given a bounded linear operator T on a Hilbert space H, we define the *essential spectrum* of T, denoted $\mathrm{Sp}_e\, T$, as the spectrum of the coset of T in the Calkin algebra $\mathcal{L}(H)/\mathcal{LC}(H)$. By Exercise III.5, we know that this Calkin algebra is semi-simple.

COROLLARY 5.3.3.   *Suppose that a bounded linear operator T on a Hilbert space is not the sum of a scalar and a compact operator. Then there exists a quasi-nilpotent operator Q such that $T + Q$ has more than one point in its essential spectrum.*

PROOF.   By Theorem 5.3.2 applied to the Calkin algebra, there exists $R \in \mathcal{L}(H)$ such that $\rho(\dot{R}) = 0$ and $\# \mathrm{Sp}_e(T + R) > 1$. By the T.T. West decomposition (see for instance S.R. Caradus, W.E. Pfaffenberger, B. Yood, *Calkin Algebras and Algebras of Operators on Banach Spaces*, New York, 1974, Theorem 5.3.2, p. 51), we can write $R = C + Q$ with C compact and Q quasi-nilpotent, and so $\# \mathrm{Sp}_e(T + Q) > 1$. $\square$

§4. Spectral Characterizations of Finite-Dimensional Banach Algebras

Let A be a Banach algebra such that $A/\mathrm{Rad}\, A$ is finite-dimensional. For all $x \in A$ the class $\dot{x}$ is algebraic in $A/\mathrm{Rad}\, A$ and consequently $\mathrm{Sp}\, x$ is finite. Surprisingly, the converse is true even supposing that the spectrum is finite on a very small part of the algebra. This result was used by I. Kaplansky to characterize ring isomorphisms of Banach algebras.

The following lemma is a generalization of the Wedderburn-Artin theorem (Theorem 2.1.2).

LEMMA 5.4.1.   *Let A be a semi-simple Banach algebra. Suppose there exists an integer $n \geq 1$ such that for all $x \in A$, x is algebraic of degree $\leq n$. Then A is the direct sum of at most n algebras isomorphic to some $M_k(\mathbf{C})$, with $k \leq n$.*

PROOF.   Let π be a continuous irreducible representation of A on a Banach space X. If $\dim X > n$, there exist $n + 1$ linearly independent vectors $\xi_1, \ldots, \xi_{n+1} \in X$. So by Theorem 4.2.5 there exists $x \in A$ such that $\pi(x)\xi_1 = \xi_1$, $\pi(x)\xi_2 = 2\xi_2, \ldots, \pi(x)\xi_{n+1} = (n+1)\xi_{n+1}$. Consequently $\{1, 2, \ldots, n+1\} \subset \mathrm{Sp}\, \pi(x) \subset \mathrm{Sp}\, x$

which is a contradiction because x is algebraic of degree n. So $k = \dim X \leq n$. Consequently, by Exercise IV.13, $\pi(A) \simeq M_k(\mathbb{C})$. Let $\pi_1, \ldots, \pi_m$ be m continuous irreducible representations of A with different kernels and let $B_m = \pi_1(A) \times \cdots \times \pi_m(A)$. For $i \neq j$, $\pi_i(\operatorname{Ker} \pi_j)$ is a two-sided ideal of $\pi_i(A) \simeq M_{k_i}(\mathbb{C})$ and consequently $\pi_i(\operatorname{Ker} \pi_j) = \pi_i(A)$ or $\pi_i(\operatorname{Ker} \pi_j) = \{0\}$. The same argument with $\pi_j(\operatorname{Ker} \pi_i)$ implies that $\pi_j(\operatorname{Ker} \pi_i) = \pi_j(A)$ or $\pi_j(\operatorname{Ker} \pi_i) = \{0\}$. Finally we obtain $A = \operatorname{Ker} \pi_i + \operatorname{Ker} \pi_j$ or $\operatorname{Ker} \pi_i = \operatorname{Ker} \pi_j$, but this last case is impossible. By the Chinese Remainder Theorem, the mapping $\phi \colon x \mapsto (\pi_1(x), \ldots, \pi_m(x))$ is onto B_m. Let $\lambda_1, \ldots, \lambda_m$ be different complex numbers. There exists $x \in A$ such that

$$\pi_1(x) = \lambda_1 1, \ldots, \pi_m(x) = \lambda_m 1.$$

But x is algebraic of degree $\leq n$ so there exists a polynomial p of degree $\leq n$ such that $p(x) = 0$. Then $p(\lambda_1) = \cdots = p(\lambda_m) = 0$ and consequently $m \leq \deg p \leq n$. If m is the greatest number $\leq n$ such that we have m continuous irreducible representations $\pi_1, \ldots, \pi_m$ with different kernels, then by Theorem 4.2.1 (ii) we have $\operatorname{Rad} A = \operatorname{Ker} \pi_1 \cap \cdots \cap \operatorname{Ker} \pi_m = \{0\}$. So ϕ is an isomorphism of A onto B_m and the lemma is proved. $\square$

In a real vector space X we say that a set U is *absorbing* if there exists $a \in U$ such that for all $x \in X$, there exists $r > 0$ such that $a + \lambda x \in U$ for $-r \leq \lambda \leq r$. For instance, an open set is absorbing but the converse is not true in general.

THEOREM 5.4.2. *Let A be a Banach algebra containing an absorbing set U such that $\operatorname{Sp} x$ is finite for all $x \in U$. Then $A/\operatorname{Rad} A$ is finite-dimensional.*

PROOF. Replacing A by $A/\operatorname{Rad} A$ and U by its image under the canonical mapping from A onto $A/\operatorname{Rad} A$, we may suppose without loss of generality that A is semi-simple. Let $a \in U$ be such that for all $x \in A$ there exists $r > 0$ such that $a + \lambda x \in U$ for $-r \leq \lambda \leq r$. Considering the analytic function $\lambda \mapsto a + \lambda(x - a) = f(\lambda)$, we have $\operatorname{Sp} f(\lambda)$ finite for λ in some real interval which has a non-zero capacity. So, by Theorem 3.4.25, $\# \operatorname{Sp}(a + \lambda(x - a)) < +\infty$ for all $\lambda \in \mathbb{C}$. In particular, $\# \operatorname{Sp} x < +\infty$ for all $x \in A$. Let $A_k = \{x \colon x \in A, \# \operatorname{Sp} x \leq k\}$. By Corollary 3.4.5, A_k is closed. So by Baire's theorem there exists a smallest integer n such that $\# \operatorname{Sp} x \leq n$ for x in a ball $B(b, s)$. Applying again the argument at the beginning of this proof, with the absorbing set $B(b, s)$, we conclude that $\# \operatorname{Sp} x \leq n$ for all $x \in A$. Let π be a continuous irreducible representation of A. The argument at the beginning of the proof of Lemma 5.4.1 implies that $\dim X \leq n$. So let $x \in A$ be arbitrary with $\operatorname{Sp} x = \{\alpha_1, \ldots, \alpha_m\}$, $m \leq n$. Suppose that $\alpha_1, \ldots, \alpha_\ell \in \operatorname{Sp} \pi(x)$, with $\ell \leq m$. By the Cayley-Hamilton theorem we have $(\pi(x) - \alpha_1 1)^n \times \cdots \times (\pi(x) - \alpha_\ell 1)^n = 0$ and consequently $(\pi(x) - \alpha_1 1)^n \times \cdots \times (\pi(x) - \alpha_m 1)^n = 0$. This being true for all such representations π, we have $(x - \alpha_1 1)^n \times \cdots \times (x - \alpha_m 1)^n = 0$ because A is semi-simple. Hence x is algebraic of degree $\leq n^2$. We now use Lemma 5.4.1 to finish the proof. $\square$

COROLLARY 5.4.3. *Let A be a Banach algebra with involution. Suppose that the real vector subspace H of self-adjoint elements contains an absorbing subset U such that $\operatorname{Sp} h$ is finite for all $h \in U$. Then $A/\operatorname{Rad} A$ is finite-dimensional.*

PROOF. As U is an absorbing set, there exists $h_0 \in U$ which satisfies the following: for $h \in H$ given, there exists $r > 0$ such that $h_0 + \lambda(h - h_0) \in U$ for $0 \leq \lambda \leq r$. By Theorem 3.4.25 we conclude that $\# \operatorname{Sp} h < +\infty$, for all $h \in H$. Now let $x = h + ik \in A$ be arbitrary, with $h, k \in H$. Considering as earlier the analytic function $\lambda \mapsto h + \lambda k$ we have $\# \operatorname{Sp}(h + \lambda k) < +\infty$, for $\lambda \in \mathbf{R}$. So $\# \operatorname{Sp}(h + \lambda k) < +\infty$ for all $\lambda \in \mathbf{C}$, and in particular for $\lambda = i$. Then by Theorem 5.4.2, $A/\operatorname{Rad} A$ is finite-dimensional. $\square$

We now give some small applications which improve a result due to R.E. Edwards.

THEOREM 5.4.4. *Let A be a Banach algebra containing a non-empty open set U of invertible elements such that $\rho(x)\rho(x^{-1}) = 1$, for all $x \in U$. Then $A/\operatorname{Rad} A$ is isomorphic to $\mathbf{C}$.*

PROOF. Let $x \in U$. There exists $r > 0$ such that $|\lambda| < r$ implies $x - \lambda 1 \in U$ and so $\rho(x - \lambda 1)\rho((x - \lambda 1)^{-1}) = 1$. By Theorem 3.3.5, we conclude that $\operatorname{Sp} x$ is included in a circle centred at λ for all $|\lambda| \leq r$. Consequently $\# \operatorname{Sp} x = 1$, for $x \in U$. The proof of Theorem 5.4.2 with $n = 1$ implies that $A/\operatorname{Rad} A \simeq \mathbf{C}$. $\square$

COROLLARY 5.4.5. *Let A be a Banach algebra containing a non-empty open set U of invertible elements such that $\|x\| \cdot \|x^{-1}\| = 1$, for all $x \in U$. Then A is isomorphic to $\mathbf{C}$.*

PROOF. Without loss of generality we may suppose that U is connected. First we prove that $\|x\| \cdot \|x^{-1}\| = 1$ on $G_1(A)$, the connected component of 1 in the set of invertible elements. Let $E = \{x : x \in G_1(A)$ such that $\|x\| \cdot \|x^{-1}\| = 1\}$. This is a closed subset of $G_1(A)$ containing 1. We now prove that it is open. Let $a \in U$ and $x \in E$. Then for $y \in Ua^{-1}$ we have: $1 \leq \|xy\| \cdot \|y^{-1}x^{-1}\| = \|xyaa^{-1}\| \cdot \|aa^{-1}y^{-1}x^{-1}\| \leq \|x\| \cdot \|ya\| \cdot \|a^{-1}\| \cdot \|a\| \cdot \|a^{-1}y^{-1}\| \cdot \|x^{-1}\| = 1$. Moreover

xUa^{-1} is a connected set containing x, so $xy \in E$ for y in the neighbourhood Ua^{-1} of the unit. Consequently $E = G_1(A)$. Suppose that $x \in \operatorname{Rad} A$, $x \neq 0$. Then $x_t = \frac{1+tx}{1+t} \in G_1(A)$ for $t \geq 0$. Hence $\|x_t\| \cdot \|x_t^{-1}\| = 1$. When t goes to $+\infty$, x_t goes to x, hence $\lim \|x_t^{-1}\| = 1/\|x\|$. By Lemma 3.2.11, x is invertible which is absurd. So $\operatorname{Rad} A = \{0\}$. Moreover we have $1 \leq \rho(x)\rho(x^{-1}) \leq \|x\| \cdot \|x^{-1}\|$ on $G_1(A)$. So by Theorem 5.4.4, A is isomorphic to $\mathbf{C}$. $\square$

§5. Automatic Continuity for Banach Algebra Morphisms

With Corollary 4.1.10 we saw that on a commutative semi-simple Banach algebra, all the Banach algebra norms are equivalent. In the 1950s, I. Kaplansky conjectured that the same result is true for non-commutative semi-simple Banach algebras. This problem was solved only in 1967 by B.E. Johnson. His proof, which is not so easy, uses mainly representation theory (see [2], pp. 128–131 or [1], pp. 161–163). Using subharmonic functions we now give a very simple proof of an extension of this result.

Let A and B be two Banach algebras and let T be a linear mapping from A into B. We define the *separating space* of T by

$$\mathfrak{S}(T) = \{a : a \in B, \; \exists(x_n) \text{ in } A, \; \lim_{n \to \infty} x_n = 0 \text{ and } \lim_{n \to \infty} Tx_n = a\}.$$

It is a closed linear subspace of B and, by the Closed Graph Theorem, T is continuous if and only if $\mathfrak{S}(T) = \{0\}$.

THEOREM 5.5.1. *Let A and B be two Banach algebras. Suppose that T is a linear mapping from A into B such that $\rho(Tx) \leq \rho(x)$ for every $x \in A$. Then $a \in \mathfrak{S}(T)$ implies $\rho(Tx) \leq \rho(a + Tx)$, for all $x \in A$. In particular $\mathfrak{S}(T) \cap T(A)$ is included in the set of quasi-nilpotent elements of B.*

PROOF. Let $a \in \mathfrak{S}(T)$ and (x_n) be such that $\lim_{n \to \infty} x_n = 0$ and $\lim_{n \to \infty} Tx_n = a$. Let $x \in A$ and $\lambda \in \mathbf{C}$ be arbitrary. Then $\lim_{n \to \infty}(\lambda x_n + x) = x$ and $\rho(T(\lambda x_n + x)) = \rho(\lambda Tx_n + Tx) \leq \rho(\lambda x_n + x)$ by hypothesis. So

$$\limsup_{n \to \infty} \rho(\lambda Tx_n + Tx) \leq \limsup_{n \to \infty} \rho(\lambda x_n + x) \leq \rho(x),$$

by upper semicontinuity of ρ on A. We set $\phi_n(\lambda) = \rho(\lambda Tx_n + Tx)$, which is subharmonic. Consequently

$$\phi(\lambda) = \limsup_{n \to \infty} \phi_n(\lambda) \leq \rho(x)$$

satisfies the mean inequality on **C**, but in general is not upper semicontinuous. We set

$$\psi(\lambda) = \limsup_{\mu \to \lambda} \phi(\mu)$$

to be its upper regularization, which is subharmonic on **C**. We have $\phi(\lambda) \leq \psi(\lambda) \leq \rho(x)$, for all $\lambda \in$ **C**. So by Liouville's theorem for subharmonic functions, ψ is constant. So $\rho(Tx) = \phi(0) \leq \psi(0) = \psi(\lambda)$ for all $\lambda \in$ **C**. By upper semicontinuity of ρ on B we have

$$\phi(\lambda) \leq \rho(\lambda a + Tx)$$

and consequently

$$\psi(\lambda) \leq \limsup_{\mu \to \lambda} \rho(\mu a + Tx) \leq \rho(\lambda a + Tx).$$

So we conclude that $\rho(Tx) \leq \rho(\lambda a + Tx)$ for all $\lambda \in$ **C**, and in particular for $\lambda = 1$. If $a \in \mathfrak{S}(T) \cap T(A)$ then $a = Tu$ for some $u \in A$. Taking $x = -u$, we get $\rho(a) = 0$, so the result. $\square$

THEOREM 5.5.2. *Suppose that we have the hypotheses of Theorem 5.5.1 with B semi-simple, and moreover that T is onto. Then T is continuous.*

PROOF. Let $a \in \mathfrak{S}(T)$ with $a = Tu$. Taking $x = y - u$, we get $\rho(Ty - a) \leq \rho(Ty)$ for all $y \in A$. So $\rho(a + q) = 0$ for all quasi-nilpotent elements q of B. By Theorem 5.3.1 (iii) we have $a \in \text{Rad } B = \{0\}$. $\square$

COROLLARY 5.5.3 (B.E. JOHNSON). *Let A and B be two Banach algebras, with B semi-simple. Suppose that T is a morphism from A onto B. Then T is continuous.*

PROOF. If T is a morphism we obviously have $\text{Sp } Tx \subset \text{Sp } x$, so $\rho(Tx) \leq \rho(x)$ for all $x \in A$. We then apply Theorem 5.5.2. $\square$

REMARK. In the proof of the corollary it is not necessary to use Theorem 5.3.1 (iii) because in this case $\mathfrak{S}(T)$ is a closed two-sided ideal of B contained in the set of quasi-nilpotent elements of B, and hence $\mathfrak{S}(T) \subset \text{Rad } B = \{0\}$.

From Corollary 5.5.3 we obtain immediately the equivalence of Banach algebra norms and the continuity of involution on a semi-simple Banach algebra (see the proofs of Corollary 4.1.10 and Corollary 4.1.11).

This last result on continuity of involution is very important for it implies that whenever we have to study a spectral problem on a Banach algebra with involution, we may suppose that the involution is continuous (we transfer the involution to $A/\operatorname{Rad} A$ which is semi-simple, and this does not modify the spectrum).

In the theory of automatic continuity, the following problem has been known for a long time and remains unsolved: if T is a morphism from a Banach algebra A into a semi-simple Banach algebra B, having a dense range in B, is it true that T is continuous?

The only partial solution we know is the following:

THEOREM 5.5.4. *Let T be a morphism from a Banach algebra A into a semi-simple Banach algebra B. Suppose that $T(A)$ is dense and has at most countable codimension in B. Then T is continuous and onto.*

PROOF. Because $T(A)$ is dense in B it is easy to verify that $\mathfrak{S}(T)$ is a two-sided closed ideal of B. Let $a \in \mathfrak{S}(T)$. Then $e^{\alpha a} - 1 \in \mathfrak{S}(T)$ for all $\alpha \in \mathbf{C}$. The cosets corresponding to these elements in the quotient linear space $B/T(A)$ are linearly dependent because $T(A)$ has at most countable codimension in B. Hence there exist $\alpha_1, \ldots, \alpha_n, \beta_1, \ldots, \beta_n$ different from 0 such that

$$u = \beta_1(e^{\alpha_1 a} - 1) + \cdots + \beta_n(e^{\alpha_n a} - 1) \in T(A).$$

So $u \in \mathfrak{S}(T) \cap T(A)$ and hence, by Theorem 5.5.1, $\rho(u) = 0$. By the Holomorphic Functional Calculus the spectrum of a is included in the set of zeros of the function

$$f(z) = \beta_1(e^{\alpha_1 z} - 1) + \cdots + \beta_n(e^{\alpha_n z} - 1),$$

which is not identically zero. So the spectrum of a is finite. By Corollary 3.4.5 the spectrum function is continuous at a. But there exists a sequence (x_n) such that $\lim_{n \to \infty} x_n = 0$ and $\lim_{n \to \infty} T x_n = a$, and so $\rho(a) = \lim_{n \to \infty} \rho(T x_n) \leq \lim_{n \to \infty} \rho(x_n) = 0$. Consequently every element of $\mathfrak{S}(T)$ is quasi-nilpotent. So $\mathfrak{S}(T) \subset \operatorname{Rad} B = \{0\}$. From a classical theorem by T. Kato, we conclude that T is onto. $\square$

For more details on these problems concerning automatic continuity, see the little book by A. Sinclair, *Automatic Continuity of Linear Operators*, Cambridge, 1976.

§6. Elements with Finite Spectrum

THEOREM 5.6.1.   *Let f be an analytic function from a domain $D \subset \mathbb{C}$ into a Banach algebra. Suppose that for every $\lambda \in D$ the element $f(\lambda)$ is algebraic. Then there exist an integer $n \geq 1$ and n holomorphic functions on D, denoted by $\alpha_1, \ldots, \alpha_n$, such that*

$$f(\lambda)^n + \alpha_1(\lambda)f(\lambda)^{n-1} + \cdots + \alpha_n(\lambda)1 = 0,$$

for all $\lambda \in D$.

PROOF.   Because $f(\lambda)$ is algebraic its spectrum is finite. By Theorem 3.4.25 there exist an integer k, a closed discrete subset E of D and k functions $h_1, \ldots, h_k$, locally holomorphic on $D \setminus E$, such that $\operatorname{Sp} f(\lambda) = \{h_1(\lambda), \ldots, h_k(\lambda)\}$ for $\lambda \in D \setminus E$, and $\# \operatorname{Sp} f(\lambda) < k$ for $\lambda \in E$. Set

$$\gamma_1(\lambda) = -(h_1(\lambda) + \cdots + h_k(\lambda))$$

$$\gamma_2(\lambda) = \sum_{1 \leq i < j \leq k} h_i(\lambda)h_j(\lambda)$$

$$\vdots$$

$$\gamma_k(\lambda) = (-1)^k h_1(\lambda) \cdots h_k(\lambda).$$

These functions are locally holomorphic on $D \setminus E$. We prove that $\gamma_1, \ldots, \gamma_k$ can be extended continuously on D. Let $\lambda_0 \in E$ and $r > 0$ be such that $\overline{B}(\lambda_0, r) \cap E = \{\lambda_0\}$. Then $\operatorname{Sp} f(\lambda_0) = \{\beta_1, \ldots, \beta_\ell\}$ for some $\ell < k$, and $\operatorname{Sp} f(\lambda) = \{h_1(\lambda), \ldots, h_k(\lambda)\}$ for $0 < |\lambda - \lambda_0| < r$. Without loss of generality we may suppose that $0 \notin \operatorname{Sp} f(\lambda_0)$ (we change $f(\lambda)$ for $f(\lambda) + \alpha 1$ with some $\alpha \in \mathbb{C}$). We then choose ℓ disjoint open disks $\Delta_1, \ldots, \Delta_\ell$, centred respectively at $\beta_1, \ldots, \beta_\ell$ and not containing 0. There exists $s < r$ such that $|\lambda - \lambda_0| < s$ implies $\operatorname{Sp} f(\lambda) \subset \Delta_1 \cup \cdots \cup \Delta_\ell$. Applying the Holomorphic Functional Calculus and Theorem 3.4.25, we conclude that there exist integers $n_1, \ldots, n_\ell \geq 1$ such that $n_1 + \cdots + n_\ell = k$ and $\#(\operatorname{Sp} f(\lambda) \cap \Delta_i) = n_i$ for $0 < |\lambda - \lambda_0| < s$. Then γ_1 can be extended continuously at λ_0 by

$$\gamma_1(\lambda_0) = -(n_1\beta_1 + \cdots + n_\ell\beta_\ell).$$

For the other symmetric functions the continuous extension is similar. Now from Morera's theorem, the k functions $\gamma_1, \ldots, \gamma_k$ are holomorphic on D. Once more by the Holomorphic Function Calculus, $g(\lambda) = f(\lambda)^k + \gamma_1(\lambda)f(\lambda)^{k-1} + \cdots + \gamma_k(\lambda)1$ is quasi-nilpotent for all $\lambda \in D$. But for each λ fixed, $g(\lambda)$ is algebraic, so $g(\lambda)$

is nilpotent. Let $D_m = \{\lambda : \lambda \in D, g(\lambda)^m = 0\}$, which is closed in D. Then $D = \cup_{m=1}^{\infty} D_m$. By Baire's theorem and the Identity Principle for analytic functions we conclude that there exists some integer $m_0 \geq 1$ such that $g(\lambda)^{m_0} = 0$ on D. Expanding this expression, we get the result. $\square$

COROLLARY 5.6.2. *Let X be a Banach space and let f be an analytic function from a domain $D \subset \mathbf{C}$ into $\mathcal{L}(X)$. Suppose that for every $\lambda \in D$ the element $f(\lambda)$ is polynomially compact. Then there exist n holomorphic functions on D, denoted by $\alpha_1, \ldots, \alpha_n$, such that*

$$f(\lambda)^n + \alpha_1(\lambda)f(\lambda)^{n-1} + \cdots + \alpha_n(\lambda)1 \in \mathcal{LC}(X),$$

for all $\lambda \in d$.

PROOF. We apply Theorem 5.6.1 to $\mathcal{L}(X)/\mathcal{LC}(X)$. $\square$

For a given Banach algebra A we denote by $\mathfrak{F}$ the set of elements of A wich have finite spectrum. This set may be extremely complicated. It contains in particular the set of quasi-nilpotent elements and the set of projections. In the rest of this section we investigate some properties of $\mathfrak{F}$.

THEOREM 5.6.3. *Suppose that $a + \lambda b \in \mathfrak{F}$ for all $\lambda \in \mathbf{C}$. Then we have $(a - \alpha 1)(b - \beta 1)^{-1} \in \mathfrak{F}$ for all $\alpha \in \mathbf{C}$ and $\beta \in \mathbf{C} \setminus \mathrm{Sp}\, b$.*

PROOF. Let $\beta \notin \mathrm{Sp}\, b$. We have $a + \lambda(b - \beta 1) \in \mathfrak{F}$ for all $\lambda \in \mathbf{C}$. But $\alpha \in \mathrm{Sp}(a + \lambda(b - \beta 1))$ is equivalent to $(a - \alpha 1)(b - \beta 1)^{-1} + \lambda 1$ non-invertible and consequently to $-\lambda \in \mathrm{Sp}((a - \alpha 1)(b - \beta 1)^{-1})$. We now apply Theorem 3.4.26 to the analytic function $f(\lambda) = a + \lambda(b - \beta 1)$. Suppose that $\{\lambda : \alpha \in \mathrm{Sp}\, f(\lambda)\}$ is closed and discrete. By the previous equivalence we conclude that $\mathrm{Sp}((a - \alpha 1)(b - \beta 1)^{-1})$ is compact and discrete, hence finite. If we have $\alpha \in \mathrm{Sp}\, f(\lambda)$ for all $\lambda \in \mathbf{C}$ then $\frac{\alpha}{\lambda} \in \mathrm{Sp}(\frac{a}{\lambda} + b - \beta 1)$ for $\lambda \neq 0$, so $0 \in \limsup_{|\lambda| \to \infty} \mathrm{Sp}(\frac{a}{\lambda} + b - \beta 1) \subset \mathrm{Sp}(b - \beta 1)$, which gives a contradiction. Hence the theorem is proved. $\square$

COROLLARY 5.6.4. *If for some $a \in A$ we have $a + \mathfrak{F} \subset \mathfrak{F}$ then $a\mathfrak{F} \subset \mathfrak{F}$.*

PROOF. Let $b \in \mathfrak{F}$. Adding a constant if necessary, we may suppose that b is invertible, and so $b^{-1} \in \mathfrak{F}$. By hypothesis $a + \lambda b^{-1} \in \mathfrak{F}$ for all $\lambda \in \mathbf{C}$. By the previous theorem with $\alpha = \beta = 0$, we get $ab = a(b^{-1})^{-1} \in \mathfrak{F}$. $\square$

LEMMA 5.6.5.   *If for some $a \in A$ we have $a + \mathfrak{F} \subset \mathfrak{F}$ then there exists an integer $n \geq 1$ such that $\# \operatorname{Sp}[a, x] \leq n$, for all $x \in A$.*

PROOF.   Clearly $a \in \mathfrak{F}$ and $e^x a e^{-x} \in \mathfrak{F}$ for all $x \in A$, and thus $e^x a e^{-x} - a \in \mathfrak{F}$. Let A_k be the set of $x \in A$ such that $\# \operatorname{Sp}(e^x a e^{-x} - a) \leq k$. We have $A = \cup_{k=1}^{\infty} A_k$ and by Corollary 3.4.5, A_k is closed. So by Baire's theorem there exists a smallest integer n such that A_n contains a ball $B(b, r)$. We fix $x \in A$ and consider

$$\phi(\lambda) = e^{b + \lambda(x - b)} a e^{-b - \lambda(x - b)} - a \in \mathfrak{F}.$$

This function ϕ is analytic, and for $|\lambda| \cdot \|x - b\| < r$ we have $\# \operatorname{Sp} \phi(\lambda) \leq n$. So, by Theorem 3.4.25, we have $\# \operatorname{Sp} \phi(\lambda) \leq n$ for all $\lambda \in \mathbf{C}$. This implies that $x \in A_n$. Hence $A = A_n$. Considering

$$f(\lambda) = \begin{cases} \dfrac{a - e^{\lambda x} a e^{-\lambda x}}{\lambda} & , \text{ for } \lambda \neq 0 \\[2mm] [a, x] & , \text{ for } \lambda = 0 \end{cases}$$

we have $\# \operatorname{Sp} f(\lambda) \leq n$, for $\lambda \neq 0$. Once more by Theorem 3.4.25 we conclude that $\# \operatorname{Sp} f(0) = \# \operatorname{Sp}[a, x] \leq n$. $\square$

THEOREM 5.6.6.   *If for some $a \in A$ we have $a + \mathfrak{F} \subset \mathfrak{F}$ then a is algebraic modulo the radical of A.*

PROOF.   By the previous lemma there exists an integer n such that $\# \operatorname{Sp}[a, x] \leq n$, for all $x \in A$. Let π be a continuous irreducible representation of A on a Banach space X. Let $\xi \in X$ be such that $\xi_0 = \xi$, $\xi_1 = \pi(a)\xi, \ldots$, $\xi_{n+1} = \pi(a)^{n+1}\xi$ are linearly independent in X. For $\alpha_0, \alpha_1, \ldots, \alpha_{n+1}$ given in X, by Theorem 4.2.5, there exists $x \in A$ such that $\pi(x)\xi_0 = \alpha_0, \ldots, \pi(x)\xi_{n+1} = \alpha_{n+1}$. So $[\pi(a), \pi(x)]\xi_i = i\xi_i$ for $i = 0, \ldots, n$, if we take $\alpha_1 = \pi(a)\alpha_0$, $\alpha_2 = \pi(a)\alpha_1 - \xi_1, \ldots$, $\alpha_{n+1} = \pi(a)\alpha_n - n\xi_n$. Hence $\{0, \ldots, n\} \subset \operatorname{Sp}[\pi(a), \pi(x)] \subset \operatorname{Sp}[a, x]$, which gives a contradiction. Consequently $\xi_1, \ldots, \xi_{n+1}$ are linearly dependent, and by Theorem 4.2.7 $\pi(a)$ is algebraic of degree $\leq n + 1$. We have $a \in \mathfrak{F}$ and so $\operatorname{Sp} a = \{\beta_1, \ldots, \beta_\ell\}$. Consequently $(\pi(a) - \beta_1 1)^{n+1} \times \cdots \times (\pi(a) - \beta_\ell 1)^{n+1} = 0$ for every continuous irreducible representation π, and so $(a - \beta_1 1)^{n+1} \times \cdots \times (a - \beta_\ell 1)^{n+1} \in \operatorname{Rad} A$. $\square$

COROLLARY 5.6.7.   *Let T be a bounded linear operator on a Hilbert space H which is not polynomially compact. Then there exists $U \in \mathcal{L}(H)$ such that $\operatorname{Sp}_e U$ is finite and $\operatorname{Sp}_e(T + U)$ is infinite.*

PROOF.   We apply Theorem 5.6.6 to the Calkin algebra $\mathcal{L}(H)/\mathcal{LC}(H)$ which is semi-simple. $\square$

THEOREM 5.6.8.   *Let A be a semi-simple Banach algebra. Suppose that $q_0 \in A$ is a non-nilpotent quasi-nilpotent element. Then there exists another quasi-nilpotent element $q_1 \in A$ such that $\mathrm{Sp}(q_0 + q_1)$ is infinite.*

PROOF.   Suppose that $q_0 - e^x q_0 e^{-x} \in \mathfrak{F}$ for all $x \in A$. The same argument as in the proof of Lemma 5.6.5 implies that $\# \mathrm{Sp}[q_0, x] \leq n$ for all $x \in A$ and some integer $n \geq 1$. As in the proof of Theorem 5.6.6 we conclude that $q_0^{n+1} = 0$, so we have a contradiction. Consequently there exists some $x \in A$ such that for $q_1 = e^x q_0 e^{-x}$, we have $\mathrm{Sp}(q_0 + q_1)$ infinite. $\square$

If $(e_n)_{n \geq 0}$ is the standard basis of $\ell^2(\mathbf{N})$, then considering the two operators a, b defined by

$$ae_n = \begin{cases} e_{n+1} & \text{, if } n \text{ is odd} \\ 0 & \text{, if } n \text{ is even} \end{cases}, \qquad be_n = \begin{cases} 0 & \text{, if } n \text{ is odd} \\ e_{n+1} & \text{, if } n \text{ is even} \end{cases}$$

we have $a^2 = b^2 = 0$ and $(a + b)e_n = e_{n+1}$ for $n \geq 0$. So $a + b$ is the unilateral shift whose spectrum is the unit circle (see [3], Problem 85). For a general Banach space X, is it possible to build two quasi-nilpotent operators whose sum has infinite spectrum? This problem is difficult because in general X has no topological basis so it is impossible to give an explicit construction. Nevertheless we shall solve the problem using a circuitous method.

LEMMA 5.6.9 (S. GRABINER).   *Let A be a Banach algebra such that its set of nilpotent elements contains a linear subspace on which the degree of nilpotency is unbounded. Then A contains a non-nilpotent quasi-nilpotent element which is a limit of nilpotent ones.*

PROOF.   Let M be the set of nilpotent elements of A. We denote by X the closure of the linear subspace contained in M on which the degree of nilpotency is unbounded. Let $E_k = \{x : x \in M \cap X, x^k = 0\}$. Then $M \cap X = \cup_{k \geq 1} E_k$ and each of the E_k is closed in X. We now show that E_k has no interior point in X. If a is interior to E_k then, for $x \in X$, we have $a + \lambda(x - a) \in E_k$ for λ small. Hence $(a + \lambda(x - a))^k = 0$ for all $\lambda \in \mathbf{C}$, by the Identity Principle. Consequently $x^k = 0$, and the degree of nilpotency is bounded on X, a contradiction.

Now let N be the set of quasi-nilpotent elements of A. Then $N = \cap_{n \geq 1}\{x : \rho(x) < 1/n\}$, so N is a G_δ-set by upper-semicontinuity of ρ. The set $M \cap X$ is dense in X so the quasi-nilpotent elements of X form a dense G_δ-subset of X. By Baire's theorem we conclude that the set of quasi-nilpotent elements of X is not a countable union of closed subsets of X with empty interior, so it is different from $M \cap X$. $\square$

THEOREM 5.6.10. *Let X be a Banach space of infinite dimension. Then there exist two quasi-nilpotent and compact operators T_1, T_2 on X such that $\mathrm{Sp}(T_1 + T_2)$ is infinite.*

PROOF. We prove that $A = \mathfrak{LC}(X) + \mathbb{C}I$ satisfies the hypothesis of Lemma 5.6.9. Let $(X_k)_{k \geq 0}$ be a sequence of finite-dimensional linear subspaces of X such that $X_0 = \{0\}$ and X_k is strictly included in X_{k+1}, for $k \geq 0$. Let $F_n = \{T : T \in \mathfrak{L}(X), T(X) \subset X_n$ and $T(X_k) \subset X_{k-1}$ for $k = 1, \ldots, n\}$ and let $F = \cup_{n \geq 1} F_n$. Then F is a linear subspace of the set of nilpotent elements of A and contains elements with degree of nilpotency as large as we want as X is infinite-dimensional. Moreover, by Exercise III.4, A is semi-simple. So, using Theorem 5.6.8, the theorem is proved. □

§7. Inessential Elements

There are many results in spectral theory concerning the relation between the spectrum of an operator and its essential spectrum, that is, the spectrum of the class of this operator in the quotient algebra obtained from the closed two-sided ideal of compact operators. These include the theorems of B.A. Barnes, I.C. Gohberg, D.C. Kleinecke and A.F. Ruston which are given below.

In this section we show that the hypothesis that the elements of the closed two-sided ideal are compact is irrelevant. The essential assertion is that these elements have a spectrum which is either finite or a sequence converging to zero. With this point of view many results in spectral theory can be extended and greatly simplified. We only present a selected list of new dishes obtained by this "nouvelle cuisine". The main ingredient in these arguments is Theorem 3.4.26.

Let I be a two-sided ideal (not necessarily closed) of a Banach algebra A. We say that I is *inessential* if, for every $x \in I$, the spectrum of x has at most 0 as a limit point. For instance in $\mathfrak{L}(X)$ the set $\mathfrak{F}$ and the set $\mathfrak{LC}(X)$ of compact operators are two-sided inessential ideals. Given a two-sided ideal I of A we denote by $\mathrm{kh}(I)$ the intersection of all kernels of continuous irreducible representations π of A such that $I \subset \mathrm{Ker}\,\pi$. It is easy to see that $I \subset \overline{I} \subset \mathrm{kh}(I)$, and that $\mathrm{kh}(I)$ is the inverse image of the radical of $A/\overline{I}$.

Let x be in A and α be isolated in the spectrum of x. We define the *projection associated to x and α* by

$$p = \frac{1}{2\pi i} \int_\Gamma (\lambda\mathbf{1} - x)^{-1}\, d\lambda$$

where Γ is a contour around α, separating α from the remaining spectrum of x. In fact p does not depend on the contour Γ, as long as Γ separates α from the rest of the spectrum. Thus we can suppose that Γ is a small circle with centre at α.

LEMMA 5.7.1. *Let I be a two-sided ideal of A and let $x \in \mathrm{kh}(I)$. Suppose that $\alpha \neq 0$ is isolated in the spectrum of x. Then the projection associated to x and α is in I.*

PROOF. Let Γ be a circle centred at α, separating α from 0 and from the rest of the spectrum. For $\lambda \in \Gamma$ we have

$$(\lambda \mathbf{1} - x)^{-1} = \frac{1}{\lambda} + \frac{1}{\lambda} x (\lambda \mathbf{1} - x)^{-1}.$$

So we have

$$p = \frac{1}{2\pi i} \int_\Gamma \frac{d\lambda}{\lambda} + \frac{x}{2\pi i} \int_\Gamma \frac{1}{\lambda} (\lambda \mathbf{1} - x)^{-1} \, d\lambda.$$

The first term is zero and the second term is in $\mathrm{kh}(I)$, so $p \in \mathrm{kh}(I)$. Let $\dot{p}$ denote the coset of p in $A/\overline{I}$. Then $\dot{p} \in \mathrm{Rad}(A/\overline{I})$ and so $\rho(\dot{p}) = 0$, where ρ denotes the spectral radius. But $\dot{p}$ is also a projection, consequently $\dot{p} = 0$, and hence $p \in \overline{I}$. Moreover $p\overline{I}p$ is a closed subalgebra of A, hence a Banach algebra with identity p, in which pIp is a dense two-sided ideal, and so $pIp = p\overline{I}p$. Then $p = p^3 \in p\overline{I}p = pIp \subset I$. $\square$

The argument shows that I and $\mathrm{kh}(I)$ have the same set of projections.

The following result is an improvement of a classical result of D.C. Kleinecke.

THEOREM 5.7.2. *Let I and J be two two-sided inessential ideals of A having the same set of projections. Denoting by $x + I$ (resp. $x + J$) the coset of x in A/I (resp. A/J), then $x + I$ is invertible in A/I if and only if $x + J$ is invertible in A/J. If moreover I and J are closed, then $\mathrm{Sp}(x + I) = \mathrm{Sp}(x + J)$, for all $x \in A$.*

PROOF. Suppose that $x + J$ is invertible in A/J but that $x + I$ is not invertible in A/I. Without loss of generality we may suppose that $x + I$ is not right invertible. Then there exists $y \in A$ such that $a = xy - 1 \in J$. If $1 + a$ is invertible then $xy(1 + a)^{-1} = 1$, and so $x + I$ is right invertible. Consequently $-1 \in \mathrm{Sp}\,a$. Because J is inessential, -1 is isolated in $\mathrm{Sp}\,a$. By Lemma 5.7.1, the corresponding projection p is in J, so by hypothesis it is also in I. By the Holomorphic Functional Calculus it is easy to see that $-1 \notin \mathrm{Sp}(a - ap)$, so that $xy - ap = 1 + a - ap$ has an inverse z in A. Consequently $(x + I)(yz + I) = 1 + I$ which is a contradiction. By a symmetric argument we get that $x + I$ is invertible if and only if $x + J$ is invertible. Replacing x by $x - \lambda \mathbf{1}$ we get the last part of the theorem. $\square$

In particular this result can be applied with $\mathfrak{F}$, $\overline{\mathfrak{F}}$ and $\mathfrak{LC}(X)$, all of which have the same set of projections. So we get the following:

COROLLARY 5.7.3 (D.C.KLEINECKE). *Let X be a Banach space and let T be a bounded linear operator on X. Then we have* $\mathrm{Sp}(T+\overline{\mathfrak{F}}) = \mathrm{Sp}(T+\mathfrak{LC}(X)) = \mathrm{Sp_e}(T)$.

We shall see below that if I is a two-sided inessential ideal then $\overline{I}$ and $\mathrm{kh}(I)$ are also inessential. Thus Theorem 5.7.2 can be used in that case.

Let I be a fixed inessential two-sided ideal of A. For x in A, we define $D(x)$ in the following way:

$$\lambda \notin D(x) \iff \begin{cases} \lambda \notin \mathrm{Sp}\,x \\ \text{or} \\ \lambda \text{ is an isolated spectral value of } x \text{ with} \\ \text{the corresponding projection in } I. \end{cases}$$

It is easy to verify that $D(x)$ is compact and that $\mathrm{Sp}\,x \setminus D(x)$ is discrete, and hence finite or countable. It is also obvious that $D(x - \lambda 1) = D(x) - \lambda$ for every $\lambda \in \mathbf{C}$.

The next result is a strong improvement of a theorem obtained previously by I.C. Gohberg for $A = \mathfrak{L}(X)$ and $I = \mathfrak{LC}(X)$ (see for instance L.C. Gohberg and M.G. Krejn, *Introduction à la théorie des opérateurs linéaires non auto-adjoints dans un espace hilbertien*, Chapter 1, Theorem 5.1 and Lemma 5.2).

THEOREM 5.7.4 (PERTURBATION BY INESSENTIAL ELEMENTS). *Let I be a two-sided inessential ideal of a Banach algebra A. For $x \in A$ and $y \in I$ we have the following properties:*

(i) *if G is a connected component of $\mathbf{C} \setminus D(x)$ intersecting $\mathbf{C} \setminus \mathrm{Sp}(x+y)$ then it is a component of $\mathbf{C} \setminus D(x+y)$,*

(ii) *the unbounded connected components of $\mathbf{C} \setminus D(x)$ and $\mathbf{C} \setminus D(x+y)$ coincide, in particular $D(x)$ and $D(x+y)$ have the same external boundaries,*

(iii) *if $\dot{x}$ denotes the coset of x in $A/\overline{I}$ then we have $\mathrm{Sp}\,\dot{x} \subset D(x)$ and $D(x)\hat{} = (\mathrm{Sp}\,\dot{x})\hat{}$, where $\hat{}$ denotes the polynomially convex hull of the set.*

PROOF. (i) Let $G' = G \setminus \mathrm{Sp}\,x$ and let $\lambda \in G'$. Then G' is a domain such that $G \setminus G'$ is discrete. We have

$$\lambda 1 - (x + y) = (\lambda 1 - x)[1 - (\lambda 1 - x)^{-1}y]. \tag{1}$$

Let $f(\lambda) = (\lambda 1 - x)^{-1} y$, which is analytic on G', and has values in I. By hypothesis and Theorem 3.4.26, we have either $1 \in \operatorname{Sp} f(\lambda)$ for all $\lambda \in G'$, or $\{\lambda : \lambda \in G', 1 \in \operatorname{Sp} f(\lambda)\}$ closed and discrete in G'. Suppose we are in the first situation. Because $G \setminus \operatorname{Sp}(x + y)$ is a non-empty open set, $G' \setminus \operatorname{Sp}(x + y)$ is non-empty. Equation (1) implies $G' \subset \operatorname{Sp}(x + y)$, so we get a contradiction. Hence for all $\lambda \in G'$ we must have $\lambda 1 - (x + y)$ invertible except on a closed discrete subset, and consequently $\lambda 1 - (x + y)$ is invertible for $\lambda \in G$ except on a discrete subset. Let α be such a point of the discrete subset of G, and suppose that $\alpha \in \operatorname{Sp} x$. Then there exists a small circle Γ, with centre at α, isolating α from the rest of the spectrum of x and from the rest of the discrete subset of G. If $\lambda \in \Gamma$, then $\lambda 1 - (x + y)$ and $\lambda 1 - x$ are invertible. Moreover we have

$$(\lambda 1 - (x + y))^{-1} = (\lambda 1 - x)^{-1} + (\lambda 1 - (x + y))^{-1} y (\lambda 1 - x)^{-1}. \qquad (2)$$

The last term of the second member is in I, and because $\alpha \in \operatorname{Sp} x \setminus D(x)$ we have

$$\frac{1}{2\pi i} \int_\Gamma (\lambda 1 - x)^{-1} \, d\lambda \in I.$$

Then by (2) and Lemma 5.7.1 we have $(\frac{1}{2\pi i}) \int_\Gamma (\lambda 1 - (x + y))^{-1} \, d\lambda \in I$ and consequently $\alpha \notin D(x + y)$. If $\alpha \notin \operatorname{Sp} x$, the same argument works except that $(\frac{1}{2\pi i}) \int_\Gamma (\lambda 1 - x)^{-1} \, d\lambda = 0$. Then $G \subset \mathbb{C} \setminus D(x + y)$. Let H be the connected component of $\mathbb{C} \setminus D(x + y)$ containing G. If $H \cap (\mathbb{C} \setminus \operatorname{Sp} x)$ is empty, then $G \subset H \subset \operatorname{Sp} x$ and $G \subset \mathbb{C} \setminus D(x)$. Thus $G \subset \operatorname{Sp} x \setminus D(x)$ which is absurd because this last set is discrete. So H intersects $\mathbb{C} \setminus \operatorname{Sp} x$ and we may apply the previous argument to H to conclude that $H \subset \mathbb{C} \setminus D(x)$, and hence that $H = G$.

(ii)   This property follows immediately from property (i) if we notice that the intersection of the unbounded components of $\mathbb{C} \setminus D(x)$ and $\mathbb{C} \setminus D(x + y)$ contains the set of z such that $|z| > \max(\|x\|, \|x + y\|)$.

(iii)   Let $\alpha \neq 0$, $\alpha \in \operatorname{Sp} x \setminus D(x)$, and let p be its associated projection, which is in I. We have

$$\operatorname{Sp} \dot{x} = \operatorname{Sp}(\overline{x - xp}) \subset \operatorname{Sp}(x - xp).$$

By the Holomorphic Functional Calculus, $\alpha \notin \operatorname{Sp}(x - xp)$, and thus $\alpha \notin \operatorname{Sp} \dot{x}$. Hence $\operatorname{Sp} \dot{x} \subset D(x) \cup \{0\}$. Taking $\lambda \notin \operatorname{Sp} \dot{x}$, we have

$$\operatorname{Sp} \dot{x} - \lambda = \operatorname{Sp}(\overline{x - \lambda 1}) \subset D(x - \lambda 1) = D(x) - \lambda,$$

and so $\operatorname{Sp} \dot{x} \subset D(x)$. In particular, we have $(\operatorname{Sp} \dot{x})\hat{\ } \subset D(x)\hat{\ }$. Denoting by ∂_e the external boundary of a set, we have $\partial_e D(x) = \partial_e D(x + y)$ and thus $\partial_e D(x) \subset \operatorname{Sp}(x + y)$, for every $y \in I$. But by upper semicontinuity of the spectrum this inclusion is also true for $y \in \bar{I}$. By Theorem 3.3.8, we have $\partial_e D(x) \subset (\operatorname{Sp} \dot{x})\hat{\ }$, hence $D(x)\hat{\ } = (\partial_e D(x))\hat{\ } \subset (\operatorname{Sp} \dot{x})\hat{\ }\hat{\ } = (\operatorname{Sp} \dot{x})\hat{\ }$ and the proof is complete. $\square$

REMARK. Let H be a Hilbert space. Taking $x \in \mathcal{L}(H)$ and $y \in \mathcal{LC}(H)$, it is false in general that $D(x) = D(x + y)$. By inessential perturbations, some holes may appear. For instance on $H = l^2(\mathbf{Z})$, taking the two weighted shifts

$$ ae_n = \begin{cases} 0 & \text{, if } n = -1 \\ e_{n+1} & \text{, if } n \neq -1 \end{cases}, \qquad be_n = \begin{cases} e_0 & \text{, if } n = -1 \\ 0 & \text{, if } n \neq -1 \end{cases} $$

we have b of rank one, and so in $\mathcal{LC}(H)$, and we have $D(a) = \mathrm{Sp}\,a = \{z\colon |z| \leq 1\}$, $D(a + b) = \mathrm{Sp}(a + b) = \{z\colon |z| = 1\}$.

In 1954, using a rather complicated argument, A.F. Ruston proved that if $T \in \mathcal{L}(X)$ has an essential spectral radius equal to zero, then the spectrum of T is either finite or a sequence converging to zero, and the projections associated to the non-zero spectral values are in $\overline{\mathfrak{F}}$. By Corollary 5.7.3, the condition that $\rho(\dot{T}) = 0$ in the Calkin algebra $\mathcal{L}(X)/\mathcal{LC}(X)$ is equivalent to saying that for every $\epsilon > 0$ there exists an integer N such that for every $n \geq N$ there exists $T_n \in \mathfrak{F}$ with $\|T^n - T_n\| < \epsilon^n$. A.F. Ruston called such an operator *asymptotically quasi-compact*. This result derives immediately from the following:

COROLLARY 5.7.5. *Let I be a two-sided inessential ideal of a Banach algebra A. Let $x \in A$ and suppose that $\rho(\dot{x}) = 0$, where $\dot{x}$ denotes the coset of x in $A/\overline{I}$. Then the spectrum of x has at most 0 as a limit point and, for every non-zero spectral value of x, the associated projection is in I.*

PROOF. If $\rho(\dot{x}) = 0$, by Theorem 5.7.4 (iii) we have $D(x)\hat{\ } = \{0\}$, so $D(x) = 0$. Hence we get the result. $\square$

COROLLARY 5.7.6. *Let I be a two-sided inessential ideal of a Banach algebra A. Then $\mathrm{kh}(I)$ is inessential, so in particular $\overline{I}$ is inessential.*

PROOF. If $x \in \mathrm{kh}(I)$ then $\dot{x} \in \mathrm{Rad}(A/\overline{I})$, so $\rho(x) = 0$. We then apply the previous corollary. $\square$

If a Banach algebra A has minimal left ideals (resp. minimal right ideals) then by definition its *socle*, denoted by $\mathrm{soc}(A)$, is the sum of the minimal left ideals (it is also equal to the sum of minimal right ideals, so it is a two-sided ideal). The reader will find more information on the socle in [1], pp. 78–87. Using a rather complicated method, B.A. Barnes proved that every element of $\mathrm{kh}(\mathrm{soc}(A))$ has at most 0 as a limit point in its spectrum. This proof was simplified by J.C. Alexander and M.R. Smyth. In fact this result derives from Corollary 5.7.6.

COROLLARY 5.7.7. *In a Banach algebra with minimal left (or right) ideals,* kh(soc(A)) *is an inessential ideal.*

PROOF. By Corollary 5.7.6 it is sufficient to prove that soc(A) is inessential. But for every $x \in$ soc(A), the spectrum of x is finite. For instance this can be seen using the fact that the algebra xAx is finite dimensional, so x is algebraic (see Exercise V.11). $\square$

If dim $A < +\infty$ then the socle of A is non-zero because $A =$ soc(A). Conversely if A is semi-simple and $A =$ soc(A) then, by Theorem 5.4.2 , A is finite-dimensional. If X is a Banach space and $A = \mathfrak{L}(X)$ then the socle is non-zero because it contains all finite-rank operators. It would be interesting to have more examples of Banach algebras with non-zero socle.

The next result was proved by B.A. Barnes (see *On the existence of minimal ideals in a Banach algebra*, Trans. AMS **133** (1968), pp. 511–517). He first obtained the commutative case using a deep result called the Šilov Idempotent Theorem (see [10], Chapter 8). In the proof given below we eliminate this argument by using a subharmonic one.

THEOREM 5.7.8 (B.A. BARNES). *Let A be a semi-simple Banach algebra such that the spectrum of every element of A is at most countable. Then* soc(A) $\neq \{0\}$.

PROOF. Suppose that soc(A) $= \{0\}$. If for all $x \in A$ we have $\# \operatorname{Sp} x = 1$ then, by Theorem 5.4.4 (or Exercise III.21), $A \simeq \mathbf{C}$, so we get a contradiction. Hence there exists $x_0 \in A$ such that $\operatorname{Sp} x_0$ contains at least two isolated points α_0, α_1. We choose two disjoint open disks D_0, D_1, of radius ≤ 1, respectively centred at α_0, α_1, and such that $\overline{D}_0 \cap \operatorname{Sp} x_0 = \{\alpha_0\}$ and $\overline{D}_1 \cap \operatorname{Sp} x_0 = \{\alpha_1\}$. Translating x_0 if necessary, we may suppose that $0 \notin \operatorname{Sp} x_0$. By Theorem 3.4.2, there exists $r_0 < 1$ such that $\|x - x_0\| < r_0$ implies $D_0 \cap \operatorname{Sp} x \neq \emptyset$, $D_1 \cap \operatorname{Sp} x \neq \emptyset$ and $\operatorname{Sp} x \cap (\partial D_0 \cup \partial D_1 \cup \{0\}) = \emptyset$. For $i = 0, 1$, let p_i be the projection associated to x_0 and α_i. We have $p_i x_0 = x_0 p_i$, $p_0 p_1 = p_1 p_0 = 0$ and $\operatorname{Sp} p_i x_0 = \{0, \alpha_i\}$. Consider the subalgebras $A_i = p_i A p_i$. They are semi-simple Banach algebras with identity p_i (see Exercise III.6). We have

$$\operatorname{Sp}_{A_i} p_i x p_i \subset \operatorname{Sp}_A p_i x p_i.$$

Consequently the spectrum is at most countable on these subalgebras. By hypothesis, x_0 is invertible, so the previous inclusions prove that $\operatorname{Sp}_{A_i} p_i x_0 p_i = \{\alpha_i\}$. Let $r < r_0/(\|p_0\|^2 + \|p_1\|^2)$ be such that $\|x - x_0\| < r$ implies $\operatorname{Sp}_{A_i} p_i x_0 p_i \subset D_i$, for $i = 0, 1$. Consider the two sets

$$G_i = \{x : \|x - x_0\| < r, \ \# \operatorname{Sp}_{A_i} p_i x p_i > 1\}.$$

These two sets are open by Theorem 3.4.4. Suppose that $B(x_0, r) \setminus G_0$ contains an open set U. Then we have $\# \operatorname{Sp}_{A_0} y = 1$, for all y in the open subset $p_0 U p_0$ of A_0. By Exercise V.4 we have $p_0 A p_0 = \mathbb{C} p_0$, so p_0 is in the socle of A by Exercise V.10, which is absurd because $p_0 \neq 0$. Consequently G_0 is dense in $B(x_0, r)$ and the same is true for G_1. In particular $G_0 \cap G_1 \neq \emptyset$. Let $y \in G_0 \cap G_1$. Then $\operatorname{Sp}_{A_i} p_i y p_i$ contains at least two isolated points $\alpha_{i0}, \alpha_{i1} \in D_i$ $(i = 0, 1)$. We set

$$x_1 = p_0 y p_0 + p_1 y p_1 + (x_0 - p_0 x_0 - p_1 x_1).$$

By Exercise III.9 we have

$$\operatorname{Sp} x_1 \supset (\operatorname{Sp} p_0 y p_0 \cup \operatorname{Sp} p_1 y p_1 \cup \operatorname{Sp}(x_0 - p_0 x_0 - p_1 x_1)) \setminus \{0\}$$
$$\supset \{\alpha_{00}, \alpha_{01}, \alpha_{10}, \alpha_{11}\}.$$

We then continue the process with the projections p_{ij} associated to x_1 and the α_{ij} $(i, j = 0, 1)$. Let $\mathfrak{D}$ be the set of finite diadic sequences, that is of finite sequences of 0s and 1s. If $s, s' \in \mathfrak{D}$, we say that $s \leq s'$ if s' is obtained from s by adding some 0s or 1s (for instance $0\,1 \leq 0\,1\,0\,1$). Then by induction on n it is possible to build a sequence (x_n) of elements of A, a sequence of radii $r_n \leq 1/2^n$, and for each $s \in \mathfrak{D}$ a disk D_s, having the following properties:

(i) $\|x_n - x_{n+1}\| \leq 1/2^n$ and $\overline{B}(x_n, r_n) \subset B(x_{n-1}, r_{n-1})$,

(ii) if $s \leq s'$ then $\overline{D}_s \subset \overline{D}_{s'}$, otherwise $\overline{D}_s$ and $\overline{D}_{s'}$ are disjoint,

(iii) $\alpha_s \in \operatorname{Sp} x_n \cap D_s$, for all s of length $n + 1$,

(iv) $\operatorname{diam}(D_s) \leq 1/2^{n-1}$ if s has length n.

By condition (a), the sequence (x_n) converges to some element $x \in A$. Now if σ is an infinite diadic sequence, we can define $\alpha(\sigma)$ to be the unique element of the intersection of all the $\overline{D}_{\sigma_n}$, where σ_n denotes the finite diadic sequence obtained from σ by taking only its first n elements. By condition (c), we have $\operatorname{Sp} x_n \cap D_{\sigma_n} \neq \emptyset$, and so by upper semicontinuity of the spectrum we have $\alpha(\sigma) \in \operatorname{Sp} x$. Moreover if $\sigma \neq \sigma'$, then for some n, $\overline{D}_{\sigma_n}$ and $\overline{D}_{\sigma'_n}$ are disjoint, and consequently $\alpha(\sigma) \neq \alpha(\sigma')$. Because the set of σ is uncountable, $\operatorname{Sp} x$ is uncountable, so we get a contradiction. Hence $\operatorname{soc}(A) \neq \{0\}$. Thus we get a contradiction and the proof is complete. $\square$

Is it possible to give the precise algebraic structure of Banach algebras for which the spectrum of every element is at most countable, at least in the separable case? The answer is yes, by Theorem 5.7.9.

In the following, $\alpha \in \Omega$ will mean that α is an ordinal of the first or second class (see W. Sierpiński, *Cardinal and Ordinal Numbers*, Warsaw, 1965).

Let A be an arbitrary Banach algebra. We take $A_0 = A/\operatorname{Rad} A$ and inductively we define $A_n = A_{n-1}/\operatorname{kh}(\operatorname{soc}(A_{n-1}))$. The corresponding morphisms of A onto $A_0, A_1, \ldots, A_n, \ldots$ are denoted by $\phi_0, \phi_1, \ldots, \phi_n \ldots$ and their kernels by $I_0 = \operatorname{Rad} A$, $I_1 = \operatorname{kh}(\operatorname{soc}(A)), \ldots, I_n = \operatorname{Ker} \phi_n, \ldots$ We then define A_ω, with $I_\omega = \operatorname{kh}(\cup_{n \geq 1} I_n)$, and ϕ_ω, where ω is the first infinite ordinal. For every $\alpha \in \Omega$ it is possible to define A_α and ϕ_α, by transfinite induction, in the following way:

— if α is not a limit ordinal, $A_\alpha = A_{\alpha-1}/\operatorname{kh}(\operatorname{soc}(A_{\alpha-1}))$ and $\phi_\alpha = \pi_{\alpha-1} \circ \phi_\alpha$, where $\pi_{\alpha-1}$ is the canonical morphism from $A_{\alpha-1}$ onto A_α,

— if α is a limit ordinal we take $I_\alpha = \operatorname{kh}(\cup_{\beta < \alpha} I_\beta)$, $A_\alpha = A/I_\alpha$ and ϕ_α the corresponding canonical morphism.

By definition we shall say that A_α is the α-*Calkin algebra* associated to A. It is easy to verify that it is semi-simple.

THEOREM 5.7.9.   *Let A be a separable Banach algebra such that the spectrum of every element of A is at most countable. Then there exist $\alpha_0 \in \Omega$ and an ordinal composition sequence $(I_\alpha)_{\alpha \leq \alpha_0}$ of closed two-sided ideals of A such that $I_0 = \operatorname{Rad} A$, $1 \leq \operatorname{codim} I_{\alpha_0+1} < +\infty$, $I_{\alpha_0+1} = A$ and $I_{\alpha+1}/I_\alpha$ is a modular annihilator algebra for $\alpha \leq \alpha_0$.*

PROOF.   The I_α are closed in the separable space A, and the family $(I_\alpha)_{\alpha \in \Omega}$ is increasing for inclusion. A classical result of topology (see C. Kuratowski, *Topologie I*, Varsovie, 1958, p. 146) implies that this family stabilizes at some $\beta \in \Omega$. If $A_\beta \neq \{0\}$, then because $\operatorname{Sp}_{A_\beta} \phi_\beta(x) \subset \operatorname{Sp} x$, the algebra A_β satisfies the hypotheses of Theorem 5.7.8. Hence $\operatorname{soc}(A_\beta) \neq \{0\}$ which implies that $I_{\beta+1} \neq I_\beta$, a contradiction. Consequently $I_\beta = A$. Then from the definition of I_β and from Corollary 3.2.2, β is not a limit ordinal. We take $\alpha_0 = \beta - 1$. Then by Exercise V.12, A_{α_0} is finite-dimensional. Moreover $I_{\alpha+1}/I_\alpha$ can be identified with $\operatorname{kh}(\operatorname{soc}(A_\alpha))$ so, by Corollary 5.7.5 and Corollary 5.7.7, every element of $I_{\alpha+1}/I_\alpha$ has its spectrum with at most 0 as a limit point. By B.A. Barnes's characterization of modular annihilator algebras and the fact that $\operatorname{Rad} \operatorname{kh}(\operatorname{soc}(A_\alpha)) = \operatorname{kh}(\operatorname{soc}(A_\alpha)) \cap \operatorname{Rad} A_\alpha = \{0\}$ (Exercise III.7), we obtain that $I_{\alpha+1}/I_\alpha$ is modular annihilator. $\square$

We do not give the definition of a modular annihilator algebra, because it is rather technical. We refer the reader to [6].

EXERCISE 1.   Let a, b be two $n \times n$ matrices such that $a(ab - ba) = 0$. Prove that $ab - ba$ is nilpotent.

EXERCISE 2.   Let A be a Banach algebra such that $\delta(x + y) \leq \delta(x) + \delta(y)$, for all $x, y \in A$. Prove that A is commutative modulo its radical.

∗EXERCISE 3.   Let A be a real Banach algebra. Suppose that ρ is subadditive or submultiplicative or uniformly continuous on A. What can be said about A?

EXERCISE 4.   Let A be a semi-simple Banach algebra. Let U be a non-empty open subset of A such that $\mathrm{Sp}\, x$ contains one point for all x in U. Prove that A is isomorphic to $\mathbf{C}$.

EXERCISE 5.   Extend Theorem 5.3.1 and Theorem 5.4.2 to a real Banach algebra.

∗EXERCISE 6.   Let A be a semi-simple Banach algebra such that for every $x \in A$ there exists $y \in A$ satisfying $x = x^2 y$.

(i) Prove that A is commutative.

(ii) Prove that A is finite-dimensional.

(iii) Conclude that A is isomorphic to $\mathbf{C}^n$, for some integer n.

∗EXERCISE 7.   Prove that a Noetherian Banach algebra is finite-dimensional.

EXERCISE 8.   Let A be a Banach algebra and let $\mathfrak{F}_n = \{x : x \in A, \# \mathrm{Sp}\, x \leq n\}$. Suppose that $a + \mathfrak{F}_n \subset \mathfrak{F}_n$, for some $a \in A$. Prove that a is algebraic of degree less than or equal to n.

EXERCISE 9.   Let X be a Banach space of infinite dimension. Prove that there exist two quasi-nilpotent and compact operators T_1, T_2 on X such that $\mathrm{Sp}(T_1 T_2)$ is infinite.

EXERCISE 10.   Let A be a Banach algebra. Prove that for every minimal left ideal (resp. right ideal) I of A there exists a projection $p \in A$ such that $I = Ap$ (resp. pA) and $pAp = \mathbf{C}p$. Such a p is called a *minimal projection*. Is the converse true?

∗EXERCISE 11.   Let $a \in \mathrm{soc}(A)$. Prove that aAa is finite-dimensional. If you do not succeed look at [1], p. 81.

EXERCISE 12.   Let A be a semi-simple Banach algebra such that $A = \mathrm{kh}(\mathrm{soc}(A))$. Prove that A is finite-dimensional.

EXERCISE 13. Let I be a two-sided ideal of a Banach algebra A. Prove that $\mathrm{kh}(\mathrm{kh}(I)) = \mathrm{kh}(I)$.

**EXERCISE 14. Extend Theorem 5.7.8 to a real Banach algebra.

***EXERCISE 15. Extend Theorem 5.7.8 to a Banach algebra with involution, supposing that the spectrum of every self-adjoint element is at most countable.

Chapter VI

REPRESENTATION THEORY
FOR C^*-ALGEBRAS
AND THE SPECTRAL THEOREM

§1. Banach Algebras with Involution

Among Banach algebras there are very interesting ones called *Banach algebras with involution*. A mapping $x \to x^*$ from a Banach algebra A into itself is called an *involution* on A if it satisfies the following properties for all $x, y \in A$ and $\lambda \in \mathbf{C}$:

(i) $(x + y)^* = x^* + y^*$,

(ii) $(\lambda x)^* = \bar{\lambda} x^*$,

(iii) $(xy)^* = y^* x^*$,

(iv) $(x^*)^* = x$.

There are many examples of such Banach algebras with involution. Among the commutative ones there are $C(K)$, the algebra of continuous functions on a compact set K, with the involution $f \to \bar{f}$; the disk algebra $A(\Delta)$, with the involution defined by $f^*(z) = \overline{f(\bar{z})}$; and the group algebra $L^1(G)$ for a locally compact commutative group G (for instance $\mathbf{R}$), with the involution $f^*(x) = \overline{f(-x)}$ (supposing that the operation on G is denoted by $+$). For non-commutative Banach algebras with involution we have $\mathcal{L}(H)$, the algebra of bounded operators on a Hilbert space H, with the standard involution; all closed sub-algebras of $\mathcal{L}(H)$ which are invariant by this involution, and so in particular $\mathcal{LC}(H)$, the algebra of compact operators on H; and also $L^1(G)$, the group algebra of a locally compact non-commutative group G, with the involution $f^*(x) = m(x^{-1})\overline{f(x^{-1})}$, m being the modular function.

Let A be a Banach algebra with involution. Then any $x \in A$ such that $x = x^*$ is called *self-adjoint*, any $x \in A$ such that $xx^* = x^*x$ is called *normal*, and any $x \in A$ such that $xx^* = x^*x = 1$ is called *unitary*.

THEOREM 6.1.1. *Let A be a Banach algebra with involution and let $x \in A$. Then*

(i) $x + x^*$, $i(x - x^*)$, xx^* and x^*x are self-adjoint,

(ii) x has a unique representation $x = h + ik$ with h, k self-adjoint given by $h = \frac{1}{2}(x + x^*)$, $k = \frac{1}{2i}(x - x^*)$,

(iii) x is normal if and only if in the decomposition $x = h + ik$, with h, k self-adjoint, we have $hk = kh$,

(iv) the unit 1 is self-adjoint and x is invertible if and only if x^* is invertible, in which case $(x^*)^{-1} = (x^{-1})^*$,

(v) $\operatorname{Sp} x^* = \{\bar{\lambda} : \lambda \in \operatorname{Sp} x\}$, in particular $\rho(x^*) = \rho(x)$,

(vi) $x \in \operatorname{Rad} A$ if and only if $x^* \in \operatorname{Rad} A$, in particular $A / \operatorname{Rad} A$ has an involution $(\dot{x})^* = (x^*)^{\cdot}$.

PROOF. Statement (i) is obvious. In (ii) we only prove the uniqueness. If $x = u + iv$, with u, v self-adjoint, then $u - h = i(k - v)$. Taking the conjugate we get $u - h = -i(k - v)$, so $k = v$ and $u = h$. If x is normal it is obvious that h and k commute. Conversely if h and k commute then $xx^* = (h + ik)(h - ik) = h^2 + k^2 = (h - ik)(h + ik) = x^*x$. Since $1^* = 1.1^*$, (i) implies that 1 is self-adjoint, and the remaining part of (iv) comes from (3) applied to x and x^{-1}. If we apply (iv) then $x - \lambda 1$ is invertible if and only if $x^* - \bar{\lambda}1$ is invertible, so we get (v). If $x \in \operatorname{Rad} A$, then by Theorem 3.1.3, $\rho(xy) = 0$ for all $y \in A$. Thus, by (v), $\rho(y^*x^*) = 0$ for all $y \in A$, which implies by (4) that $\rho(zx^*) = 0$ for all $z \in A$. Once again by Theorem 3.1.3, we conclude that $x^* \in \operatorname{Rad} A$. If $\dot{x}$ denotes the coset of x in $A / \operatorname{Rad} A$ then $(\dot{x})^* = (x^*)^{\cdot}$ is well-defined by (v) and it is easy to verify that it is an involution on $A / \operatorname{Rad} A$. □

As a consequence of Theorem 5.5.2 we have the following important fact.

THEOREM 6.1.2 (B.E.JOHNSON). *Let A a semi-simple Banach algebra with involution. Then the involution is continuous on A.*

LEMMA 6.1.3 (P.CIVIN-B.YOOD). *Let A be a Banach algebra with involution and x be normal in A. Then there exists a closed and commutative subalgebra B, containing x, stable by involution, such that $\operatorname{Sp}_B x = \operatorname{Sp}_A x$.*

PROOF. Let $\mathfrak{C}$ be the family of subsets E of A such that $a, b \in E \cup E^*$ implies $ab = ba$. The family $\mathfrak{C}$ contains $\{x\}$ and is inductive for the order defined by inclusion. So, by Zorn's lemma, there exists a maximal set $B \in \mathfrak{C}$ containing x. Because B is maximal we have $B \cup B^* = B$ so $B = B^*$. A similar argument with $B+B$, $C \cdot B$ and $B \cdot B$ shows that B is a commutative subalgebra of A. We now prove that B is closed. If $a, b \in \overline{B} \cup \overline{B}^*$ then $a = \lim a_n$, $b = \lim b_n$, with $a_n, b_n \in B \cup B^*$, but $a_n b_n = b_n a_n$ and so $ab = ba$. Obviously we have $\mathrm{Sp}_A\, x \subset \mathrm{Sp}_B\, x$. Conversely if $\lambda \in \mathrm{Sp}_B\, x$ and $\lambda \notin \mathrm{Sp}_A\, x$, then $x - \lambda 1$ is invertible in A with $(x - \lambda 1)^{-1} \notin B$, but $B \cup \{(x - \lambda 1)^{-1}\} \in \mathfrak{C}$. So we get a contradiction. Hence $\mathrm{Sp}_A\, x = \mathrm{Sp}_B\, x$. $\square$

THEOREM 6.1.4. *Let A be a Banach algebra with involution. Suppose that $x \in A$ is a self-adjoint element such that its spectrum contains no real number $\lambda \leq 0$. Then there exists a self-adjoint element $y \in A$ such that $x = y^2$.*

PROOF. By Lemma 6.1.3 there exists a closed and commutative subalgebra B of A, stable by involution, such that $\mathrm{Sp}_A\, x = \mathrm{Sp}_B\, x$, so if we succeed in proving the theorem for B, the proof will be finished. In other words, without loss of generality we may suppose that A is commutative. Let D be the complement in C of the set of real numbers $\lambda \leq 0$. There exists f holomorphic on D such that $f^2(z) = z$ and $f(1) = 1$. Since $\mathrm{Sp}\, x \subset D$ we may define $y = f(x)$ which satisfies $y^2 = x$ by the Holomorphic Functional Calculus. We now prove that $y = y^*$. Since D is simply connected, by Runge's theorem (see [7], Chapter 13) there exists a sequence (p_n) of polynomials converging to f uniformly on each compact subset of D. Let q_n be the polynomials with real coefficients defined by

$$2q_n(z) = p_n(z) + \overline{p_n(\bar{z})}.$$

Since $f(\bar{z}) = \overline{f(z)}$, the polynomials q_n converge to f, uniformly on each compact subset of D. Let $y_n = q_n(x)$. Then $y_n = y_n^*$. By Theorem 3.3.3 (iv), $y = \lim_{n \to \infty} y_n$. If the involution is continuous then the proof is complete. Suppose not: then, by Theorem 6.1.2 applied to $A/\mathrm{Rad}\, A$, we conclude that $\dot{y} = \dot{y}^*$, so $y - y^* \in \mathrm{Rad}\, A$. By Theorem 6.1.1 we have $y = h + ik$, with h, k self-adjoint, so $k \in \mathrm{Rad}\, A$. The algebra A being supposed commutative, we have $x = h^2 - k^2 + 2ihk$ so $hk = 0$. But $0 \notin \mathrm{Sp}\, x$ and $k^2 \in \mathrm{Rad}\, A$ implies h^2 invertible, so h is invertible. Hence we have $k = h^{-1}(hk) = 0$. $\square$

§2. C^*-algebras

If $f \in C(K)$, then $\|f\bar{f}\|_\infty = \|f\|_\infty^2$. If A is a closed subalgebra of $\mathfrak{L}(H)$, stable by involution, for some Hilbert space H, then by Theorem 2.3.1 (iii) we have $\|xx^*\| = \|x\|^2$, for all $x \in A$. These examples suggest the introduction of a particular class of Banach algebras with involution called C^*-algebras. By definition they are those A such that $\|xx^*\| = \|x\|^2$, for all $x \in A$.

Of course there are examples of Banach algebras with involution which are not C^*-algebras. For instance, $A(\Delta)$ with the involution $f^*(z) = \overline{f(\bar{z})}$ and $L^1(\mathbf{R})$ with the involution $f^*(x) = \overline{f(-x)}$ (see Exercise VI.1).

THEOREM 6.2.1. *Let A be a C^*-algebra. Then we have the following properties:*

(i) $\|x\| = \|x^*\|$, *for all $x \in A$, so the involution is continuous,*

(ii) *if h is self-adjoint then e^{ih} is unitary,*

(iii) *if h is self-adjoint then $\operatorname{Sp} h \subset \mathbf{R}$,*

(iv) $\rho(xx^*) = \|x\|^2$ *for all $x \in A$, in particular for all x normal we have $\rho(x) = \|x\|$,*

(v) *if x, y are normal then $\triangle(\operatorname{Sp} x, \operatorname{Sp} y) \leq \|x - y\|$, consequently the spectrum function is uniformly continuous on the set of normal elements.*

PROOF. We have $\|x\|^2 = \|xx^*\| \leq \|x\| \|x^*\|$ so $\|x\| \leq \|x^*\|$. Consequently $\|x^*\| \leq \|(x^*)^*\| = \|x\|$, hence (i) is proved. This implies in particular that the involution is continuous. If $h = h^*$ we have $e^{ih} = \lim_{n \to \infty} p_n(ih)$, where $p_n(z) = 1 + z + \frac{z^2}{2!} + \cdots + \frac{z^n}{n!}$. Consequently $p_n(ih)^* = p_n(-ih)$, so $(e^{ih})^* = \lim_{n \to \infty} p_n(-ih) = e^{-ih}$. This implies (ii). We now prove (iii). Suppose that $\alpha + i\beta \in \operatorname{Sp} h$, with $\beta \neq 0$. Taking $k = \frac{1}{\beta}(h - \alpha\mathbf{1})$ we have $k = k^*$ and $i \in \operatorname{Sp} k$, consequently $1 + ik$ is not invertible in A. Given any $\lambda \in \mathbf{R}$ we now have $(\lambda + 1)\mathbf{1} - (\lambda\mathbf{1} - ik)$ not invertible in A so, in particular, $|\lambda + 1| \leq \|\lambda\mathbf{1} - ik\|$. Consequently $(\lambda + 1)^2 \leq \|\lambda\mathbf{1} - ik\|^2 = \|(\lambda\mathbf{1} - ik)(\lambda\mathbf{1} - ik)^*\| = \|\lambda^2 + k^2\| \leq \lambda^2 + \|k\|^2$. So we get $2\lambda \leq \|k^2\| - 1$, for all $\lambda \in \mathbf{R}$, and this gives a contradiction. Hence (iii) is proved. Suppose that $h = h^*$. Then we have $\|h^2\| = \|h\|^2$. By induction it is easy to prove that $\|h^{2^n}\| = \|h\|^{2^n}$. Consequently, using Theorem 3.2.8, we get $\rho(h) = \lim_{n \to \infty} \|h^{2^n}\|^{1/2^n}$. Taking $h = xx^*$ we get $\rho(xx^*) = \|xx^*\| = \|x\|^2$. If x and x^* commute, by Corollary 3.2.10, we have $\|x\|^2 = \rho(xx^*) \leq \rho(x)\rho(x^*) = \rho(x)^2 \leq \|x\|^2$. So $\rho(x) = \|x\|$. To finish we prove (v). Suppose that there exists $\lambda \in \operatorname{Sp} y$ such that $\operatorname{dist}(\lambda, \operatorname{Sp} x) > \|x - y\|$. By Theorem 3.3.5, we have $\operatorname{dist}(\lambda, \operatorname{Sp} x) = 1/\rho((\lambda\mathbf{1} - x)^{-1})$. But $(\lambda\mathbf{1} - x)^{-1}$ is normal so, by (iv), we have $\operatorname{dist}(\lambda, \operatorname{Sp} x) = 1/\|(\lambda\mathbf{1} - x)^{-1}\|$. Consequently $\|(\lambda\mathbf{1} - x)^{-1}(x - y)\| \leq \|(\lambda\mathbf{1} - x)^{-1}\| \|(x - y)\| < 1$, so

$$\lambda\mathbf{1} - y = (\lambda\mathbf{1} - x)[1 + (\lambda\mathbf{1} - x)^{-1}(x - y)]$$

is invertible, which is a contradiction. Interchanging x and y we conclude that $\triangle(\operatorname{Sp} x, \operatorname{Sp} y) \leq \|x - y\|$. $\square$

REMARK. Property (iv) of Theorem 6.2.1 implies the uniqueness of the C^*-algebra norm.

COROLLARY 6.2.2. *Let A be a commutative C^*-algebra and let $x \in A$. For every character χ we have $\chi(x^*) = \overline{\chi(x)}$.*

PROOF. Let $x = h + ik$, with h, k self-adjoint. If χ is a character then $\chi(h) \in \operatorname{Sp} h$ and $\chi(k) \in \operatorname{Sp} k$, so by property (iii) of Theorem 6.2.1, $\chi(h)$ and $\chi(k)$ are real numbers. Consequently $\chi(x^*) = \overline{\chi(x)}$. $\square$

In the situation of a C^*-algebra, Lemma 6.1.3 can be improved.

COROLLARY 6.2.3. *Let A be a C^*-algebra and let B be a closed subalgebra of A stable by involution, containing the unit 1. Then for every $x \in B$ we have $\operatorname{Sp}_A x = \operatorname{Sp}_B x$.*

PROOF. Obviously we have $\operatorname{Sp}_A x \subset \operatorname{Sp}_B x$. So suppose that $x \in B$ and x is invertible in A. Then we must show that x is invertible in B. Now x^* and xx^* are also invertible in A. By Theorem 6.2.1 (iii), we have $\operatorname{Sp}_A(xx^*) \subset \mathbf{R} \setminus \{0\}$. Consequently $\mathbf{C} \setminus \operatorname{Sp}_A(xx^*)$ is connected. By Corollary 3.2.14, $\operatorname{Sp}_A(xx^*) = \operatorname{Sp}_B(xx^*)$, so xx^* is invertible in B. Hence $x^{-1} = x^*(xx^*)^{-1} \in B$. $\square$

COROLLARY 6.2.4 (V.PTÁK-J.ZEMÁNEK). *Let $M = (a_{ij})$ be a normal $n \times n$ matrix and let r be the square root of $\sum_{j=2}^{n} |a_{1j}|^2 = \sum_{i=2}^{n} |a_{i1}|^2$. Then there exists an eigenvalue λ of M such that $|a_{11} - \lambda| \leq r$.*

PROOF. Let P be the projection having zero entries, except on the first line and the first column where each entry is 1, and let $Q = I - P$. Define $N = PMP + QMQ$. We have $\|M - N\| = r$ so, by property (v) of Theorem 6.2.1, we have $\Delta(\operatorname{Sp} M, \operatorname{Sp} N) \leq r$. But $a_{11} \in \operatorname{Sp} N$, so we get the result. $\square$

Let x, y be two elements of a Banach algebra with involution A. Supposing that $xy = yx$, is it true that $x^*y = yx^*$? In general the answer is negative. But if A is a C^*-algebra and if x is normal then it is true. This result can even be slightly generalized by the following result.

THEOREM 6.2.5 (B.FUGLEDE-C.R.PUTNAM-M.ROSENBLUM). *Let x, y, z be elements of a C^*-algebra A. Assume that x, y are normal and that $xz = zy$. Then $x^*z = zy^*$.*

PROOF. From the hypothesis we conclude that $x^n z = zy^n$ for all $n \geq 1$. So $e^{\lambda x} z = z e^{\lambda y}$ or $z = e^{-\lambda x} z e^{\lambda y}$, for all $\lambda \in \mathbb{C}$. Consequently $e^{\lambda x^*} z e^{-\lambda y^*} = e^{\lambda x^* - \lambda x} z e^{\lambda y - \lambda y^*}$ because x and y are normal. But by continuity of the involution, the two elements $u_1 = e^{\lambda x^* - \lambda x}$ and $u_2 = e^{\lambda y - \lambda y^*}$ are unitary. So $\|u_1\| = \|u_2\| = 1$. Considering the analytic function $f(\lambda) = e^{\lambda x^*} z e^{-\lambda y^*}$ we have $\|f(\lambda)\| \leq \|u_1 z u_2\| \leq \|z\|$. So by Liouville's theorem, $f(\lambda)$ is constant and equal to z. Hence $e^{\lambda x^*} z = z e^{\lambda y^*}$ for all $\lambda \in \mathbb{C}$ and this implies $x^* z = z y^*$. □

The next theorem is one of the most important results in spectral theory. It is in fact the key to the proof of the spectral theorem for normal operators on a Hilbert space. Surprisingly it says that the only commutative C^*-algebras (with unit) are the algebras $C(K)$.

THEOREM 6.2.6 (I.M.GELFAND-M.A.NAĬMARK). *Let A be a commutative C^*-algebra and let $\mathfrak{M}$ be its set of characters. Then A is isometrically isomorphic to $C(\mathfrak{M})$.*

PROOF. By Corollary 6.2.2 we have $(x^*)\hat{} = (x\hat{})^-$. Then the Gelfand transform $x \to x\hat{}$ is a $\star$-morphism from A into $C(\mathfrak{M})$. For any $x \in A$ we have $\|x\| = \rho(x) = \|\hat{x}\|_\infty$ by Theorem 6.2.1 (iv) and Theorem 4.1.8. So $x \to \hat{x}$ is an isometry. In particular it is injective. Because the Gelfand transform is an isometry, $\hat{A}$ is complete for $\|\cdot\|_\infty$, so it is closed in $C(\mathfrak{M})$. It is also a subalgebra of $C(\mathfrak{M})$, stable by conjugation, containing the constants (because $\hat{1} = 1$) and separating the points of $\mathfrak{M}$. So by the Stone-Weierstrass theorem we conclude that $\hat{A} = C(\mathfrak{M})$. □

If A is a C^*-algebra without unit then $\mathfrak{M}$ is locally compact, but not compact. In that case A is isometrically isomorphic to $C_0(\mathfrak{M})$, the algebra of continuous functions on $\mathfrak{M}$ which tend to zero at infinity (see Exercise VI.3).

Given an element x of a Banach algebra and f holomorphic on a neighbourhood of the spectrum of x we have seen in Theorem 3.3.3 that it is possible to define $f(x)$ such that $\text{Sp } f(x) = f(\text{Sp } x)$. If x is normal in a C^*-algebra then it is possible to extend this holomorphic functional calculus to a continuous one. Even more, in §3 we shall extend it to bounded Borel functions.

THEOREM 6.2.7 (CONTINUOUS FUNCTIONAL CALCULUS). *Let A be a C^*-algebra and let x be normal in A. For every function f continuous on $\text{Sp } x$ it is possible to define a normal element $f(x)$ in A such that we have the following properties:*

(i) *the mapping $f \to f(x)$ is an algebra morphism from $C(\text{Sp } x)$ into A such that $1(x) = 1$, $I(x) = x$ (where $I(\lambda) = \lambda$) and $f(x)^* = \bar{f}(x)$,*

(ii) Sp $f(x) = f(\text{Sp } x)$, so in particular $\|f(x)\| = \sup\{|f(\lambda)| : \lambda \in \text{Sp } x\}$,

(iii) if (f_n) converges uniformly to f in $C(\text{Sp } x)$ then $f_n(x)$ converges to $f(x)$.

PROOF.   Let B be the closed and commutative subalgebra of A generated by 1, x, x^*. It is stable by involution, so by Theorem 6.2.6, it is isometrically isomorphic to $C(\mathfrak{M})$, where $\mathfrak{M}$ denotes its set of characters. We now show that $\mathfrak{M}$ can be identified with $\text{Sp}_A x$. Let Φ be the mapping from $\mathfrak{M}$ into $\text{Sp}_B x$ defined by

$$\Phi : \chi \to \chi(x).$$

By Theorem 4.1.2, it is surjective. If $\chi_1(x) = \chi_2(x)$, then by Corollary 6.2.2 $\chi_1(x^*) = \chi_2(x^*)$, so by continuity χ_1 and χ_2 coincide on all B, that is $\chi_1 = \chi_2$. By definition of the Gelfand topology, the mapping Φ is a continuous bijection from $\mathfrak{M}$ onto $\text{Sp}_B x$. By Corollary 6.2.3, $\text{Sp}_A x = \text{Sp}_B x$. Since $\mathfrak{M}$ is compact it is homeomorphic to $\text{Sp}_A x$. Let Ψ denote the isometric isomorphism from $C(\text{Sp } x)$ onto B. For $f \in C(\text{Sp } x)$ we define

$$f(x) = \Psi(f).$$

Property (i) is obvious. By Corollary 6.2.3 we have $\text{Sp}_A f(x) = \text{Sp}_B f(x) = f(\text{Sp}_B x) = f(\text{Sp}_A x)$. Consequently $\|f(x)\| = \rho(f(x)) = \sup\{|f(\lambda)| : \lambda \in \text{Sp } x\}$. Property (iii) follows from $\|f_n(x) - f(x)\| = \sup\{|f_n(\lambda) - f(\lambda)| : \lambda \in \text{Sp } x\}$. $\square$

If m is a normal $n \times n$ matrix then it is well-known that it is diagonalizable and that the eigenspaces associated to different eigenvalues are orthogonal. In other words, if $\{\lambda_1, \ldots, \lambda_k\}$ are the different eigenvalues of m, we have $m = \sum_{i=1}^{k} \lambda_i p_i$, where the p_i are self-adjoint orthogonal projections, that is they verify $p_i^2 = p_i$, $p_i^* = p_i$, for all $1 \leq i \leq k$, $p_i p_j = 0$ for $i \neq j$ and $p_1 + \cdots + p_k = I$. In fact this result follows from Theorem 6.2.7.

COROLLARY 6.2.8.   *Let A be a C^*-algebra and let x be a normal element of A having a finite spectrum $\{\lambda_1, \ldots, \lambda_k\}$. Then there exist self-adjoint orthogonal projections $p_1, \ldots, p_k$ in the commutative closed subalgebra generated by 1, x, x^* such that $p_1 + \cdots + p_k = 1$ and $x = \sum_{i=1}^{k} \lambda_i p_i$.*

PROOF.   On $\text{Sp } x = \{\lambda_1, \ldots, \lambda_k\}$ we define the k functions $\chi_1, \ldots, \chi_k$ by

$$\chi_i(\lambda) = \begin{cases} 1 & \text{if } \lambda = \lambda_i \\ 0 & \text{if } \lambda \neq \lambda_i \, . \end{cases}$$

Obviously we have $\chi_i^2 = \chi_i$, $\chi_i = \overline{\chi_i}$, $\chi_i \chi_j = 0$ for $i \neq j$, $\chi_1 + \cdots + \chi_k = 1$ and $\sum_{i=1}^{k} \lambda_i \chi_i = I$, where $I(\lambda) = \lambda$, on $\text{Sp } x$. Then we apply Theorem 6.2.7 to the $p_i = \chi_i(x)$. $\square$

An element x of a C^*-algebra A is said to be *positive*, denoted $x \geq 0$, if it is self-adjoint and if its spectrum contains only positive real numbers. For instance if h is self-adjoint then by Theorem 6.2.1 (iii) we have $\operatorname{Sp} h^2 \subset \{\lambda^2 : \lambda \in \mathbf{R}\}$, so $h^2 \geq 0$.

The positive elements of a C^*-algebra play an important rôle which will be illustrated by several of the following theorems.

THEOREM 6.2.9. *Let A be a C^*-algebra and let h be a self-adjoint element of A. Then there exists a unique decompositon $h = u - v$ such that $u \geq 0, v \geq 0$ and $uv = vu = 0$.*

PROOF. By Theorem 6.2.1 we have $\operatorname{Sp} h \subset \mathbf{R}$. On the real line the function $f(t) = t$ can be written $f = f^+ - f^-$ where $f^+(t) = \max(t, 0)$, $f^-(t) = \max(-t, 0)$ and we have $f^+ \cdot f^- = 0$. Then we apply Theorem 6.2.7 to h and the three functions f, f^+, f^- to prove the existence of u and v. If we have another decomposition $h = r - s$ with $r \geq 0$, $s \geq 0$ and $rs = sr = 0$, then $hr = rh$ and $hs = sh$, so r and s commute with all the elements in the C^*-algebra generated by 1 and h. This implies that u, v, r, s commute in pairs. Let C be a commutative C^*-algebra containing $1, u, v, r, s$. Applying the Gelfand transform to C we have $\hat{u} - \hat{v} = \hat{r} - \hat{s}$, $\hat{u}\hat{v} = \hat{r}\hat{s} = 0$ and $\hat{u}, \hat{v}, \hat{r}, \hat{s} \geq 0$. This is only possible if $\hat{u} = \hat{r}$, $\hat{v} = \hat{s}$, so $u = r$ and $v = s$. □

THEOREM 6.2.10 (I.M.GELFAND-M.A.NAĬMARK). *Let A be a C^*-algebra and let $x \geq 0$ in A. Then there exists a unique $y \in A$ such that $y^2 = x$ and $y \geq 0$. Moreover y commutes with all the elements that commute with x.*

PROOF. We have $\operatorname{Sp} x \subset [0, +\infty[$. We apply Theorem 6.2.7 to $f(\lambda) = \lambda^{1/2}$. Then $f(x)$ is self-adjoint and satisfies $f(x)^2 = x$ by property (i). Moreover $\operatorname{Sp} f(x) \subset [0, +\infty[$ so $y = f(x) \geq 0$. If $ax = xa$, then by induction $ap(x) = p(x)a$ for all polynomials p, so a commutes with all elements of the subalgebra generated by 1 and x, in particular with y. We now prove the uniqueness of y. Suppose that $z^2 = x$ with $z \geq 0$. Then $zx = z^3 = xz$ so, by the previous argument, y and z commute. Let C be the closed subalgebra generated by $1, y, z$. It is a commutative C^*-algebra containing x. Applying the Gelfand transform to C we conclude that $\hat{y} = \hat{z}$ so $y = z$. □

If $x \geq 0$ this unique element y is called the *square root* of x and is denoted by $x^{1/2}$.

THEOREM 6.2.11 (M.FUKAMIYA-I.KAPLANSKY). *Let A be a C^*-algebra. Then we have the following properties:*

(i) if $u, v \in A$ with $u \geq 0$ and $v \geq 0$ then $u + v \geq 0$,

(ii) if $x \in A$ then $xx^* \geq 0$,

(iii) if $x \in A$ then $1 + xx^*$ is invertible.

PROOF. Suppose $u \geq 0$ and $v \geq 0$. Then $u + v$ is self-adjoint so, by Theorem 6.2.1 (iii), we have $\mathrm{Sp}(u + v) \subset \mathbf{R}$. Let $\alpha = \|u\|$, $\beta = \|v\|$, so that $\mathrm{Sp}\, u \subset [0, \alpha]$ and $\mathrm{Sp}\, v \subset [0, \beta]$. Consequently we have

$$\mathrm{Sp}(\alpha 1 - u) \subset [0, \alpha] \text{ and } \mathrm{Sp}(\beta 1 - v) \subset [0, \beta].$$

So we get $\|\alpha 1 - u\| = \rho(\alpha 1 - u) \leq \alpha$ and $\|\beta 1 - v\| = \rho(\beta 1 - v) \leq \beta$. Hence $\|(\alpha + \beta)1 - (u + v)\| \leq \alpha + \beta$. But we know that $\mathrm{Sp}(u + v) \subset \mathbf{R}$ so, by the previous inequality, $\mathrm{Sp}(u + v)$ is contained in the closed disk of radius $\alpha + \beta$ with centre at $\alpha + \beta$ and hence $u + v \geq 0$. We now prove (ii). Let $y = xx^*$ which is self-adjoint and let B be chosen as in Lemma 6.1.3 applied to y. By Theorem 6.2.6, B is isometrically isomorphic to $C(\mathfrak{M})$ and of course we have $\mathrm{Sp}_A\, y = \mathrm{Sp}_B\, y = \{\chi(y) : \chi \in \mathfrak{M}\}$. Then $\hat{y}$ is a real function on $\mathfrak{M}$. We have to show that $\hat{y} \geq 0$ on $\mathfrak{M}$. Since $\hat{B} = C(\mathfrak{M})$ there exists $z \in B$ such that

$$\hat{z} = |\hat{y}| - \hat{y} \text{ on } \mathfrak{M}. \tag{1}$$

Because $\hat{z}$ is real, by Theorem 6.2.7, $z = z^*$. Let $u = zx = h + ik$ with h, k self-adjoint. Then

$$uu^* = zxx^*z^* = zyz = z^2 y \tag{2}$$

$$u^*u = 2h^2 + 2k^2 - uu^* = 2h^2 + 2k^2 - z^2 y. \tag{3}$$

But h, k are self-adjoint so their spectra are real, consequently $h^2 \geq 0$ and $k^2 \geq 0$. Because $\hat{y}$ is real we have $(z^2 y)^{\hat{}} \leq 0$ on $\mathfrak{M}$. Since $z^2 y \in B$ it follows that $-z^2 y \geq 0$. So (3) and (1) imply that $u^*u \geq 0$. By Lemma 3.1.2, we have $\mathrm{Sp}(uu^*) \subset \mathrm{Sp}(u^*u) \cup \{0\}$, so $uu^* \geq 0$. This implies that $(z^2 y)^{\hat{}} \geq 0$ on $\mathfrak{M}$. With the previous argument we conclude that $(z^2 y)^{\hat{}} = 0$ on $\mathfrak{M}$, and this is possible only if $\hat{y} = |\hat{y}|$ by (1). So $y \geq 0$. Property (iii) is obvious because $-1 \notin \mathrm{Sp}(xx^*)$. $\square$

Theorem 6.2.11 implies in particular that $(xx^*)^{1/2}$ is well-defined for all $x \in A$.

COROLLARY 6.2.12 (POLAR DECOMPOSITION OF AN INVERTIBLE ELEMENT).
Let A be a C^*-algebra and let x be invertible in A. Then there exists a unique
decomposition $x = hu$ with $h \geq 0$ and u unitary. Moreover we have $h = (xx^*)^{1/2}$.

PROOF. If x is invertible, so is xx^*, hence $\mathrm{Sp}(xx^*) \subset]0, +\infty[$. Consequently, by
Theorem 6.2.7 (ii), $(xx^*)^{1/2}$ is invertible. Let $u = (xx^*)^{-1/2}x$. It is also invertible
and $uu^* = (xx^*)^{-1/2}xx^*(xx^*)^{-1/2} = 1$. So $u^{-1} = u^*$ and u is unitary. If $x = kv$
with $k \geq 0$ and v unitary then $xx^* = kvv^*k = k^2$, so by Theorem 6.2.10, we have
$k = (xx^*)^{1/2}$ and $v = u$. $\square$

We now give a very beautiful application of this last result which implies a
classical result due to B. Russo and H.A. Dye.

THEOREM 6.2.13 (L.T.GARDNER). Let A be a C^*-algebra. Denote by U the set
of unitary elements of A. If $\|x\| < 1$ there exist an integer $n \geq 2$ and n elements
$u_1, \ldots, u_n$ in U such that $x = \frac{1}{n}(u_1 + \ldots + u_n)$.

PROOF. We fix $u \in U$ and take $\|a\| < 1$. First we prove that there exist $u_1, u_2 \in U$
such that $y = \frac{a+u}{2} = \frac{u_1+u_2}{2}$. We have $y = \frac{u}{2}(1 + u^{-1}a)$ and $\|u^{-1}a\| \leq \|a\| < 1$.
So y is invertible and $\|y\| < 1$. By Corollary 6.2.12 we have $y = (yy^*)^{1/2}v$, with v
unitary. By the Holomorphic Functional Calculus $w = (yy^*)^{1/2} + i(1 - yy^*)^{1/2}$ is
well-defined, and moreover $ww^* = w^*w = 1$ and $(yy^*)^{1/2} = \frac{w+w^*}{2}$. The product of
two unitary elements being unitary, the first assumption is proved. We now consider
the sequence defined by

$$y_0 = u, \quad y_{k+1} = \frac{a + y_k}{2}.$$

Using the first part, by induction, it is easy to prove that there exist 2^k unitary
elements $u_1, \ldots, u_{2^k}$ such that

$$y_k = \frac{1}{2^k}(u_1 + \cdots + u_{2^k}).$$

Suppose now that $\|x\| < 1$. We set $a = x + \frac{1}{2^k-1}(x - u)$ where k is the smallest
integer such that $\|a\| < 1$. Then we have $x = y_k$. So the result is proved. $\square$

COROLLARY 6.2.14 (B.RUSSO-H.A.DYE). In a C^*-algebra the open unit ball is
included in the convex hull of U.

If A is finite-dimensional then it is possible to prove that $\|x\| \leq 1$ implies
$x = \frac{1}{2}(u_1 + u_2)$, for some $u_1, u_2 \in U$ (see Exercise VI.7).

If h is self-adjoint it is obvious that e^{ih} is unitary. But in general the converse is not true, even for a commutative C^*-algebra (see Exercise VI.8). If E denotes the set of e^{ih}, with h self-adjoint, it is possible to improve Theorem 6.2.13 and Corollary 6.2.14, replacing U by E (see Exercise VI.9).

With Theorem 6.2.6 we have explicitly characterized commutative C^*-algebras. Is it possible to do the same for non-commutative ones? We finish this section with this problem. But in order to solve it we need some preliminaries.

Let A be a C^*-algebra. We say that a linear functional f on A is *positive on A* if $f(xx^*) \geq 0$, for all $x \in A$. If A is commutative, so isomorphic to $C(\mathfrak{M})$, positive linear functionals correspond to positive measures on $\mathfrak{M}$. If $A = \mathcal{L}(H)$, for some Hilbert space H, then $f(T) = (T\xi|\xi)$, for some $\xi \in H$, is obviously positive. With Theorem 6.2.17 we shall see that there are many positive linear functionals on A.

THEOREM 6.2.15. *Let A be a C^*-algebra and let f be a positive functional on A. Then we have the following properties:*

(i) $f(x^*) = \overline{f(x)}$, for all $x \in A$,

(ii) $|f(xy^*)|^2 \leq f(xx^*)f(yy^*)$, for all $x, y \in A$,

(iii) $|f(x)|^2 \leq f(1)f(xx^*) \leq f(1)^2\rho(xx^*)$, in particular $|f(x)| \leq f(1)\|x\|$, for all $x \in A$, and this implies that $\|f\| = f(1)$.

PROOF. Let $x, y \in A$ and $\lambda \in \mathbb{C}$. Since $f((x + \lambda y)(x + \lambda y)^*) \geq 0$ we have

$$f(xx^*) + \lambda f(yx^*) + \bar{\lambda} f(xy^*) + \lambda\bar{\lambda} f(yy^*) \geq 0,$$

for all $\lambda \in \mathbb{C}$. Taking $\lambda = 1$ and $\lambda = i$ we conclude that $f(yx^*) = \overline{f(xy^*)}$. So with $y = 1$ we get (i). If $f(xy^*) = 0$ then (ii) is obvious. If $f(xy^*) \neq 0$, take $\lambda = tf(xy^*)/|f(xy^*)|$ with $t \in \mathbf{R}$. Then we have $f(xx^*) + 2t|f(xy^*)| + t^2 f(yy^*) \geq 0$ for all $t \in \mathbf{R}$. So we get (ii). Taking $y = 1$ in (ii) we obtain $|f(x)|^2 \leq f(1)f(xx^*)$. If $t > \rho(xx^*)$ then by Theorem 6.1.4 there exists a self-adjoint element h such that $t - xx^* = h^2$, so $tf(1) - f(xx^*) \geq 0$. This being true for all $t > \rho(xx^*)$, we have $f(xx^*) \leq f(1)\rho(xx^*)$. The rest of (iii) comes from Theorem 6.2.1 (iv). $\square$

In fact positive linear functionals can be characterized very easily.

THEOREM 6.2.16. *Let A be a C^*-algebra. A linear functional f on A is positive if and only if it is bounded and $\|f\| = f(1)$.*

PROOF. The first part is proved in Theorem 6.2.15 (iii). Suppose that $\|f\| = f(1)$. We want to prove that f is positive. First we show that $f(h) \in \mathbf{R}$ if h is self-adjoint. Let $f(h) = \alpha + i\beta$, with $\alpha, \beta \in \mathbf{R}$. We set $u = t1 - ih$, for $t \in \mathbf{R}$. Then we have

$$\|u\|^2 = \|uu^*\| = \|t^2 + h^2\| \le t^2 + \|h\|^2 .$$

We also have $|f(u)|^2 = |t - i\alpha + \beta|^2 = t^2 + \alpha^2 + \beta^2 + 2\beta t$. Hence

$$\|u\|^2 \le |f(u)|^2 - 2\beta t + \|h\|^2 \le \|u\|^2 - 2\beta t + \|h\|^2 .$$

So $2\beta t \le \|h\|^2$, for all $t \in \mathbf{R}$, and hence $\beta = 0$. We must now prove that $f(h) \ge 0$ for $h = xx^*$. Without loss of generality we may assume that $\|h\| \le 1$ and $f(1) = 1$. Then Sp $h \subset [0, 1]$ and this implies $\|1 - h\| = \rho(1 - h) \le 1$. Also $f(h) = 1 - f(1 - h) \ge 1 - \|1 - h\| \ge 0$. This completes the proof. $\square$

We say that f is a *state* on A if f is a positive linear functional, and moreover $f(1) = 1$. By the previous theorem, f is a state if and only if it is a bounded linear functional satisfying $\|f\| = f(1) = 1$. If $A = C(\mathfrak{M})$, states correspond to probability measures on $\mathfrak{M}$. Let $\mathfrak{S}$ denote the set of states of A. It is easy to see that it is a closed convex subset of the unit ball of the topological dual A' of A. So by Theorem 1.1.8 it is weakly compact. By Theorem 1.1.9, we conclude that it is the weak closure of the set of extreme points of $\mathfrak{S}$. First we show that $\mathfrak{S}$ is large, and this implies the existence of extreme states.

THEOREM 6.2.17. *Let A be a C^*-algebra and let B be a C^*-subalgebra of A containing the unit. Suppose that σ is a state on B. Then there exists a state s on A extending σ. In particular if $x \in A$ there exists an extreme state s_0 on A such that $s_0(x^*x) = \|x\|^2$.*

PROOF. On B the linear functional σ satisfies $\|\sigma\| = \sigma(1) = 1$. By the Hahn-Banach theorem there exists $s \in A'$ such that $\|s\| = \|\sigma\|$ and s extends σ on all A. So, by Theorem 6.2.16, s is a state. If $x \in A$, let B be the C^*-subalgebra generated by 1 and x^*x. By Theorem 4.1.2 and compactness of the spectrum there exists a character χ on B such that $\chi(x^*x) = \rho(x^*x) = \|x\|^2$. But χ is a state on B so, by the first part, there exists a state s on A such that $s(x^*x) = \|x\|^2$. Let Σ be the set of states s on A such that $s(x^*x) = \|x\|^2$. It is a convex weakly compact subset of the unit ball of A' so, by Theorem 1.1.9, it contains an extreme element s_0. Suppose that $s_0 = \frac{s_1 + s_2}{2}$ with $s_1, s_2 \in \mathfrak{S}$. Then we have

$$\|x\|^2 = s_0(x^*x) = \frac{1}{2}(s_1(x^*x) + s_2(x^*x)) \le \|x^*x\| = \|x\|^2 ,$$

so necessarily we have $s_1(x^*x) = s_2(x^*x) = \|x\|^2$ that is to say $s_1, s_2 \in \Sigma$, so $s_0 = s_1 = s_2$, because s_0 is extreme in Σ. Hence s_0 is extreme in $\mathfrak{S}$. $\square$

For commutative C^*-algebras, that is $C(\mathfrak{M})$, extreme states correspond to probability measures concentrated at a point, so they are the characters of A.

LEMMA 6.2.18. *Let A be a C^*-algebra and let s be a state of A. Then $N_s = \{x : x \in A, s(x^*x) = 0\}$ is a closed left ideal of A and there exists an inner product on A/N_s defined by*

$$(x + N_s | y + N_s) = s(y^*x), \text{ for all } x, y \in A,$$

such that A/N_s is a pre-Hilbertian space for that inner product.

PROOF. If $x \in N_s$ then, by Theorem 6.2.15 (ii), we have $|s(zx)|^2 \leq s(x^*x)s(zz^*)$ and so $s(zx) = 0$, for all $z \in A$. In particular $s((yx)^*(yx)) = s((x^*y^*y)x) = 0$, hence $yx \in N_s$, for all $y \in A$. Then N_s is a left ideal of A. It is closed because s is continuous. Because N_s is a left ideal the inner product is well-defined on A/N_s. The rest of the proof is left to the reader as an exercise. $\square$

COROLLARY 6.2.19. *Let A be a C^*-algebra. Then there exist a Hilbert space H and a morphism T from A into $\mathfrak{L}(H)$, such that $T(1) = I$, $T(x^*) = T(x)^*$ and $\|T(x)\| \leq \|x\|$, for all $x \in A$.*

PROOF. Let s be a state of A and define T on A, with values in $\mathfrak{L}(A/N_s)$, by

$$T(x)(y + N_s) = xy + N_s.$$

It is easy to verify that T is a morphism from A into $\mathfrak{L}(A/N_s)$ and obviously we have $T(1) = I$. We have $(T(x)(y+N_s)|T(x)(y+N_s)) = (xy+N_s|xy+N_s) = s(y^*x^*xy)$. But for $y \in A$ fixed, if we define $t(x) = s(y^*xy)$, it is a positive linear functional on A so, by Theorem 6.2.15 (iii), we have $t(x^*x) \leq t(1)\|x\|^2 = s(y^*y)\|x\|^2 = (y + N_s|y + N_s)\|x\|^2$. Hence $\|T(x)\| \leq \|x\|$, for all $x \in A$. This inequality enables us to extend the operators $T(x)$ to bounded linear operators on the completion H of A/N_s, for the norm defined by the inner product. $\square$

THEOREM 6.2.20 (I.M.GELFAND-M.A.NAÏMARK). *Let A be a C^*-algebra. Then there exist a Hilbert space H and an isometric $\star$-morphism from A onto a closed self-adjoint subalgebra of $\mathfrak{L}(H)$.*

PROOF. Let s be a state of A. We define H_s, the completion of A/N_s, and T_s, the $\star$-morphism from A into $\mathfrak{L}(H_s)$, as in Corollary 6.2.19. Let H be the Hilbert space which is the direct sum of the $(H_s)_{s\in\mathfrak{S}}$. Let $a \in A$. For any $x = (x_s)_{s\in\mathfrak{S}} \in H$ we have

$$\sum_{s\in\mathfrak{S}} \|T_s(a)(x_s)\|^2 \leq \sum_{s\in\mathfrak{S}} \|T_s(a)\|^2 \|x_s\|^2 \leq \|a\|^2 \sum_{s\in\mathfrak{S}} \|x_s\|^2 < +\infty,$$

so we can define an element $T(a)(x)$ in H such that $(T(a)(x))_s = T_s(a)(x_s)$. We have obtained a mapping $T(a)$ from A into H. It is obvious that $T(a)$ is linear, and from the above inequalities it is bounded, so $T(a) \in \mathfrak{L}(H)$ with $\|T(a)\| \leq \|a\|$. Because the T_s are $\star$-morphisms, we verify easily that T is also a $\star$-morphism from A into $\mathfrak{L}(H)$. By Theorem 6.2.17 there exists some $s \in \mathfrak{S}$ such that $s(a^\star a) = \|a\|^2$ so $\|T_s(a)\|^2 \geq (T_s(a)(1 + N_s)|T_s(a)(1 + N_s)) = s(a^\star a) = \|a\|^2$. Consequently $\|T(a)\| = \|a\|$. Therefore T is an isometry. Consequently $T(A)$ is a closed self-adjoint subalgebra of $\mathfrak{L}(H)$. Obviously $T(1) = I$. $\square$

The Hilbert space H obtained in the previous theorem is extremely large in general. But if A is separable we may suppose that $H = l^2$ (see Exercise VI.12). Theorem 6.2.20 is also true if we suppose A without unit (see Exercise VI.4).

Theorem 6.2.20 is very important because it says that a C^*-algebra can always be considered as a closed self-adjoint subalgebra of $\mathfrak{L}(H)$ for some suitable Hilbert space H.

§3. The Spectral Theorem

Throughout this section we suppose that the C^*-algebra under consideration is $\mathfrak{L}(H)$ or a closed self-adjoint subalgebra of $\mathfrak{L}(H)$ containing the identity, for a suitable Hilbert space H. Given a normal operator T we intend to extend Theorem 6.2.7 for all bounded Borel functions on $\mathrm{Sp}\, T$. First we need some definitions and preliminary results.

Let $\mathfrak{M}$ be the σ-algebra of all Borel subsets of a compact set K. We define a *resolution of the identity* to be a mapping $P : \mathfrak{M} \to \mathfrak{L}(H)$ with the following properties:

(i) $P(\emptyset) = 0$, $P(K) = I$,

(ii) each $P(U)$ is a self-adjoint projection, for $U \in \mathfrak{M}$,

(iii) $P(U \cap V) = P(U)P(V)$, for $U, V \in \mathfrak{M}$,

(iv) if $U \cap V = \emptyset$, then $P(U \cup V) = P(U) + P(V)$, for $U, V \in \mathfrak{M}$,

(v) for every $x, y \in H$ the set function $P_{x,y}$, defined by $P_{x,y}(U) = (P(U)x|y)$, is a complex regular Borel measure on $\mathfrak{M}$.

To get a very simple example we take $H = L^2([0, 1])$, $\mathfrak{M}$ the σ-algebra of Borel subsets of $[0, 1]$, and we define P by

$$P(U)f = \chi_U \cdot f,$$

for $U \in \mathfrak{M}$, $f \in L^2([0, 1])$.

It is easy to verify that a resolution of the identity P has the following properties:

(a) any two of the projections $P(U)$ commute with each other,

(b) if U and V are disjoint the associated projections are orthogonal,

(c) if $x \in H$ then $P_{x,x}$ is a positive measure of total variation $\|x\|^2$,

(d) by (iv), P is finitely additive, and if U is the countable union of disjoint U_n then the series $\sum_{n=1}^{\infty} P(U_n)$ cannot converge in norm to $P(U)$ because the norm of the projections $P(U_n)$ is one, so that in general P is not countably additive,

(e) by (v), for a fixed $x \in U$, the function $U \to P(U)x$ is a countably additive H-valued measure on $\mathfrak{M}$,

(f) if $U_n \in \mathfrak{M}$ and $P(U_n) = 0$, for $n = 1, 2, \ldots$, and if $U = \bigcup_{n=1}^{\infty} U_n$, then $P(U) = 0$.

Given an essentially bounded Borel function on K we now intend to give meaning to the integral

$$\int_K f \, dP.$$

Let P be a resolution of the identity and let f be a complex Borel function on K. Using the fact that there exists a countable base for the topology of the complex plane and using property (f) we conclude that there exists a largest open set V in $\mathbf{C}$ such that $P(f^{-1}(V)) = 0$. By definition the *essential range* of f is the complement of V. We say that f is *essentially bounded* if its essential range is bounded, hence compact. In that case we denote by $\|f\|_{\infty}$ the largest value of $|\lambda|$, for λ in the essential range. We denote by $L^{\infty}(P)$ the Banach algebra of essentially bounded complex Borel functions on K, with norm $\| \cdot \|_{\infty}$, and the convention that two functions f and g coincide if $\|f - g\|_{\infty} = 0$.

THEOREM 6.3.1. *Let P be a resolution of the identity. Then the formula*

$$(\Phi(f)x|y) = \int_K f \, dP_{x,y}, \quad \text{for } x, y \in H,$$

defines an isometric isomorphism Φ of the Banach algebra $L^\infty(P)$ onto a closed commutative self-adjoint subalgebra of $\mathfrak{L}(H)$ which contains the identity. This isomorphism has the following properties:

(i) $\Phi(1)$ is the identity of $\mathfrak{L}(H)$,

(ii) $\Phi(f)^* = \Phi(\bar{f})$, for $f \in L^\infty(P)$,

(iii) $\|\Phi(f)x\|^2 = \int_K |f|^2 \, dP_{x,x}$, for $x \in H$ and $f \in L^\infty(P)$,

(iv) $S \in \mathfrak{L}(H)$ commutes with every $P(U)$ if and only if S commutes with every $\Phi(f)$.

PROOF. Property (i) is obvious. Let $U_1, \ldots, U_n$ be a partition of K, with $U_i \in \mathfrak{M}$, and let s be a simple function such that $s = \alpha_i$ on U_i. We define $\Phi(s) \in \mathfrak{L}(H)$ by

$$\Phi(s) = \sum_{i=1}^n \alpha_i P(U_i). \tag{1}$$

Since each $P(U_i)$ is self-adjoint we have $\Phi(s)^* = \Phi(\bar{s})$. Let $V_1, \ldots, V_m$ be another similar partition and let t be a simple function such that $t = \beta_j$ on V_j. Then

$$\Phi(s)\Phi(t) = \sum_{i,j} \alpha_i \beta_j P(U_i) P(V_j) = \sum_{i,j} \alpha_i \beta_j P(U_i \cap V_j) = \Phi(st). \tag{2}$$

An analogous argument shows that Φ is linear. If $x, y \in H$, (1) leads to

$$(\Phi(s)x|y) = \sum_{i=1}^n \alpha_i (P(U_i)x|y) = \sum_{i=1}^n \alpha_i P_{x,y}(U_i) = \int_K s \, dP_{x,y}.$$

So we get

$$\|\Phi(s)x\|^2 = (\Phi(s)^*\Phi(s)x|x) = (\Phi(\bar{s}s)x|x) = \int_K |s|^2 \, dP_{x,x}.$$

Consequently

$$\|\Phi(s)x\| \leq \|s\|_\infty \|x\|, \tag{3}$$

by property (c) of the resolution of the identity. On the other hand, let i be chosen such that $|\alpha_i| = \|s\|_\infty$ and let x be in the range of $P(U_i)$. Then

$$\Phi(s)x = \alpha_i P(U_i)x = \alpha_i x \tag{4}$$

because the projections have orthogonal ranges. It follows from (3) and (4) that

$$\|\Phi(s)\| = \|s\|_\infty . \tag{5}$$

Now suppose $f \in L^\infty(P)$. There exists a sequence of simple measurable functions s_k that converges to f in the norm of $L^\infty(P)$. By (5), the corresponding operators $\Phi(s_k)$ form a Cauchy sequence in $\mathfrak{L}(H)$, so they converge in norm to an operator that we call $\Phi(f)$. It is easy to see that $\Phi(f)$ does not depend on the particular choice of the s_k. Then (5) is also true for $f \in L^\infty(P)$. Properties (ii) and (iii), being true for simple functions, also hold with $f \in L^\infty(P)$. Finally if S commutes with every $P(U)$, it commutes with $\Phi(s)$ whenever s is simple, and consequently, by continuity, with every member of $\Phi(L^\infty(P))$. $\square$

By convention we set

$$\Phi(f) = \int_K f \, dP .$$

THEOREM 6.3.2 (SPECTRAL THEOREM). *Let A be a closed, commutative self-adjoint subalgebra of $\mathfrak{L}(H)$ which contains the identity and let K be the set of characters of A. Then we have the following properties:*

(i) *there exists a unique resolution of the identity P on the Borel subsets of K such that*

$$T = \int_K \hat{T} \, dP$$

for every $T \in A$, where $\hat{T}$ is the Gelfand transform of T,

(ii) *$P(U) \neq 0$ for every non-empty open subset U of K,*

(iii) *$S \in \mathfrak{L}(H)$ commutes with every $T \in A$ if and only if S commutes with every projection $P(U)$.*

PROOF. By Theorem 6.2.6, $T \to \hat{T}$ is an isometric $\star$-isomorphism from A onto $C(K)$. Let $x, y \in H$ be given. Then $\hat{T} \to (Tx|y)$ is a bounded linear functional on $C(K)$, of norm less than or equal than $\|x\| \, \|y\|$. By Theorem 1.1.7, there exists a unique regular complex measure $\mu_{x,y}$ on K such that

$$(Tx|y) = \int_K \hat{T} \, d\mu_{x,y} , \text{ for } T \in A. \tag{6}$$

When $\hat{T}$ is real, T is self-adjoint, hence $\mu_{x,y} = \bar{\mu}_{y,x}$. By (6) and the uniqueness of $\mu_{x,y}$ we conclude that $\mu_{x,y}(U)$ is, for every Borel set $U \subset K$, a sesquilinear functional. Since $\|\mu_{x,y}\| \leq \|x\| \, \|y\|$, it follows that

$$\int_K f \, d\mu_{x,y} \tag{7}$$

is a bounded sesquilinear functional on H, for every bounded Borel function f on K. Consequently there exists an operator $\Psi(f) \in \mathcal{L}(H)$ such that

$$(\Psi(f)x|y) = \int_K f \, d\mu_{x,y}, \text{ for } x, y \in H. \tag{8}$$

By (6) we have $\Psi(\hat{T}) = T$, so Ψ is an extension of the mapping $\hat{T} \to T$ from $C(K)$ onto A. If f is real then $(\Psi(f)x|y)$ is the complex conjugate of $(\Psi(f)y|x)$, so $\Psi(f)$ is self-adjoint.

We now prove that $\Psi(fg) = \Psi(f)\Psi(g)$, for bounded Borel functions f and g. If $S, T \in A$ then $(ST)\hat{} = S\hat{}T\hat{}$. So we have

$$\int_K \hat{S}\hat{T} \, d\mu_{x,y} = (STx|y) = \int_K \hat{S} \, d\mu_{Tx,y}. \tag{9}$$

Since $\hat{A} = C(K)$ we have $\hat{T} \, d\mu_{x,y} = \mu_{Tx,y}$. Consequently formula (9) remains true if we replace $\hat{S}$ by f. Then

$$\int_K f\hat{T} \, d\mu_{x,y} = \int_K f \, d\mu_{Tx,y} = (\Psi(f)Tx|y) = (Tx|z) = \int_K \hat{T} \, d\mu_{x,z} \tag{10}$$

where $z = \Psi(f)^*y$. Once again we can replace $\hat{T}$ by g in this formula so the assertion is proved.

We now define the resolution of the identity P. If U is a Borel subset of K we set

$$P(U) = \Psi(\chi_U). \tag{11}$$

It is easy to verify that this is a resolution of the identity. Moreover $(P(U)x|y) = \mu_{x,y}(U)$ so, by (6), we have

$$T = \int_K \hat{T} \, dP. \tag{12}$$

We now prove the uniqueness of P. The regularity of the complex Borel measures $P_{x,y}$ shows that each $P_{x,y}$ is uniquely determined by (6). This follows from the uniqueness assertion in Theorem 1.1.7. Since $(P(U)x|y) = P_{x,y}(U)$, each projection $P(U)$ is also uniquely determined by (6). So the proof of (i) is complete. Suppose now that $P(U) = 0$, for U open. If $T \in A$ and $\hat{T}$ has its support in U then, by (12), we have $T = 0$, and hence $\hat{T} = 0$. By Urysohn's lemma, U must be empty. So (ii) is proved. We choose $S \in \mathcal{L}(H)$, $x, y \in H$ and we put $z = S^*y$. For any $T \in A$ and any Borel subset U of K we have

$$(STx|y) = (Tx|z) = \int_K \hat{T} \, dP_{x,z} \,, \tag{13}$$

$$(TSx|y) = \int_K \hat{T} \, dP_{Sx,y} \,, \tag{14}$$

$$(SP(U)x|y) = (P(U)x|z) = P_{x,z}(U) \,, \tag{15}$$

$$(P(U)Sx|y) = P_{Sx,y}(U) \,. \tag{16}$$

If $ST = TS$ for every $T \in A$, the measures in (13) and (14) are equal, so that $SP(U) = P(U)S$, by (15) and (16). The same argument proves the converse. This completes the proof. $\square$

The reader must keep in mind that, in general, the projections $P(U)$ are outside the algebra A except, of course, if $A = \mathcal{L}(H)$.

It is a classical result in matrix theory that a commuting family of self-adjoint $n \times n$ matrices can be simultaneously diagonalized. It is easy to see that this result is a corollary of Theorem 6.3.2. A much more important theorem is the following.

THEOREM 6.3.3 (SPECTRAL THEOREM FOR NORMAL OPERATORS). *Let $T \in \mathcal{L}(H)$ be a normal operator. Then there exists a unique resolution of the identity P on the Borel subsets of $\operatorname{Sp} T$ which satisfies*

$$T = \int_{\operatorname{Sp} T} \lambda \, dP(\lambda). \tag{17}$$

Furthermore every projection $P(U)$ commutes with every $S \in \mathcal{L}(H)$ which commutes with T.

PROOF. Let A be the smallest closed subalgebra of $\mathcal{L}(H)$ generated by I, T, T^*. It is a commutative self-adjoint subalgebra of $\mathcal{L}(H)$. By Corollary 6.2.3, $\operatorname{Sp}_A T$ is equal to $\operatorname{Sp} T$, the spectrum of T corresponding to $\mathcal{L}(H)$. So, by Theorem 6.2.6, A is isometrically isomorphic to $C(\operatorname{Sp} T)$. The existence of P follows from Theorem 6.3.2. We now prove the uniqueness. If P exists and satisfies (17), by Theorem 6.3.1, we have

$$p(T, T^*) = \int_{\operatorname{Sp} T} p(\lambda, \bar{\lambda}) \, dP(\lambda) \tag{18}$$

for an arbitrary polynomial in two variables with complex coefficients. By the Stone-Weierstrass theorem, these polynomials are dense in $C(\operatorname{Sp} T)$. Consequently the projections $P(U)$ are uniquely determined by the integrals (18), hence by T. If $TS = ST$ then, by Theorem 6.2.5, we have $T^*S = ST^*$, so S commutes with all the elements of A. Consequently S commutes with the projections $P(U)$. $\square$

We shall refer to this P as the *spectral decomposition* of T.

The previous arguments can be summarized in order to extend the Continuous Functional Calculus to bounded Borel functions.

THEOREM 6.3.4 (SYMBOLIC CALCULUS FOR NORMAL OPERATORS). *Let $T \in \mathcal{L}(H)$ be a normal operator. If f is a bounded Borel function on $\mathrm{Sp}\, T$ we set*

$$f(T) = \int_{\mathrm{Sp}\, T} f \, dP \,,$$

where P is the unique resolution of the identity associated to T by Theorem 6.3.3. The mapping $f \to f(T)$ has the following properties:

(i) *it is an algebra morphism from the algebra of all bounded Borel functions on $\mathrm{Sp}\, T$ into $\mathcal{L}(H)$ such that $1(T) = I$, $I(T) = T$ (where $I(\lambda) = \lambda$ on $\mathrm{Sp}\, T$) and $f(T)^\star = \bar{f}(T)$,*

(ii) *$\|f(T)\| \le \sup\{|f(\lambda)| : \lambda \in \mathrm{Sp}\, T\}$, the equality being true if $f \in C(\mathrm{Sp}\, T)$,*

(iii) *if f_n converges uniformly to f, then $f_n(T)$ converges in norm to $f(T)$,*

(iv) *if $S \in \mathcal{L}(H)$ and $TS = ST$, then $f(T)S = Sf(T)$ for every bounded Borel function f,*

(v) *T is the limit in norm of linear combinations of projections $P(U)$.*

If the normal operators T sit in some closed self-adjoint subalgebra A of $\mathcal{L}(H)$ we know, by Theorem 6.2.7, that for f continuous, the $f(T)$ are normal operators in A. But if f is a bounded Borel function we can only say, by Theorem 6.3.4, that the $f(T)$ are normal operators in the *bicommutant* A'' of A, that is the set of operators commuting with all the operators commuting with A. In general A'' is much larger than A.

§4. Applications

It is not true that every $T \in \mathcal{L}(H)$ has a polar decomposition (See Exercise VI.18). But we extend Corollary 6.2.12 to arbitrary normal operators.

THEOREM 6.4.1. *Let $T \in \mathcal{L}(H)$ be normal. Then it has a polar decomposition $T = PU$ where P is positive, U is unitary, and P and U commute with each other and with T.*

PROOF. Let p and u be the two bounded Borel functions on $\mathrm{Sp}\, T$ defined by $p(\lambda) = |\lambda|, u(\lambda) = \frac{\lambda}{|\lambda|}$ if $\lambda \ne 0$ and $u(0) = 1$. Using Theorem 6.3.4 we put $P = p(T)$

and $U = u(T)$. By construction P is self-adjoint. The relations $\lambda\bar{\lambda} = |\lambda|^2$, $u\bar{u} = 1$ and $\lambda = p(\lambda)u(\lambda)$ imply that $P^2 = TT^*$, consequently $P = (TT^*)^{1/2}$, $UU^* = I$ and $T = PU$. So the result is proved. $\square$

THEOREM 6.4.2. *Let $U \in \mathfrak{L}(H)$ be unitary. There exists $Q \in \mathfrak{L}(H)$, self-adjoint, such that $U = e^{iQ}$.*

PROOF. Since U is unitary, its spectrum lies on the unit circle. Consequently there exists a real bounded Borel function f on $\operatorname{Sp} U$ that satisfies $\exp(if(\lambda)) = \lambda$, for $\lambda \in \operatorname{Sp} U$. We apply Theorem 6.3.4 and take $Q = f(U)$. Obviously Q is self-adjoint and verifies $U = e^{iQ}$. $\square$

COROLLARY 6.4.3. *The group of all invertible operators in $\mathfrak{L}(H)$ is connected and every invertible operator is the product of two exponentials.*

PROOF. Let T be invertible in $\mathfrak{L}(H)$. By Corollary 6.2.12, we have $T = PU$, where $P = (TT^*)^{1/2}$ is invertible and U is unitary. Because $\operatorname{Sp} P \subset \,]0, +\infty[$, the logarithm is a continuous real function on $\operatorname{Sp} P$, and so by Theorem 6.2.7 we have $P = e^S$, for some self-adjoint $S \in \mathfrak{L}(H)$. By Theorem 6.4.2, we have $U = e^{iQ}$ for some self-adjoint $Q \in \mathfrak{L}(H)$. So $T = e^S e^{iQ}$. The continuous function $t \to e^{tS} e^{itQ}$, with $0 \le t \le 1$, defines a continuous arc from I to T. So the group of invertible elements is connected. $\square$

It is natural to ask whether every invertible $T \in \mathfrak{L}(H)$ is an exponential or, equivalently, if the product of two exponentials is an exponential. This is true for finite-dimensional algebras (see Remark following Theorem 3.3.6), but false in general (see [8], pp.317-319).

A related problem is to know if the set of exponentials is open in general. A negative answer follows from J.B. Conway and B.B. Morrel, *Roots and Logarithms of Bounded Operators on Hilbert Space*, Journal of Functional Analysis **70** (1987), pp.171-193, where they prove that T is in the interior of the set of exponentials if and only if 0 is in the unbounded component of $\operatorname{Sp} T$.

LEMMA 6.4.4. *Let A be a C^*-algebra and let I be a left ideal of A (resp. right ideal). Denote by Λ the set of finite subsets of I, ordered by inclusion. There exists a family $(u_\alpha)_{\alpha \in \Lambda}$ such that*

(i) $u_\alpha \in I$, $u_\alpha \ge 0$, $\|u_\alpha\| \le 1$,

(ii) $\lim_\alpha x u_\alpha = x$, for each $x \in \bar{I}$ (resp. $\lim_\alpha u_\alpha x = x$).

PROOF. For $\alpha = \{x_1, \ldots, x_n\} \in \Lambda$ we put

$$v_\alpha = x_1^* x_1 + \cdots + x_n^* x_n \text{ and } u_\alpha = v_\alpha \left(v_\alpha + \frac{1}{n} \right)^{-1}.$$

Because the real function $t \to t\left(t + \frac{1}{n}\right)^{-1}$ takes its values between 0 and 1, for $t \geq 0$, we have $u_\alpha \geq 0$ and $0 \leq u_\alpha \leq 1$, so (i) is proved. We have

$$\sum_{i=1}^n (x_i(u_\alpha - 1))^*(x_i(u_\alpha - 1)) = (u_\alpha - 1)v_\alpha(u_\alpha - 1) = \frac{1}{n^2} v_\alpha \left(\frac{v_\alpha + 1}{n} \right)^{-2}.$$

The real function $t \to t\left(t + \frac{1}{n}\right)^{-2}$ is less than or equal to $\frac{n}{4}$, for $t \geq 0$, consequently

$$(x_i(u_\alpha - 1))^*(x_i(u_\alpha - 1)) \leq \sum_{i=1}^n (x_i(u_\alpha - 1))^*(x_i(u_\alpha - 1)) \leq \frac{1}{4n}.$$

So we have $\|x_i(u_\alpha - 1)\|^2 \leq \frac{1}{4n}$. Consequently we have $\lim_\alpha x u_\alpha = x$, for each $x \in I$, and this is also true for the elements of $\bar{I}$. $\square$

THEOREM 6.4.5. Let A be a C^*-algebra and let I be a closed two-sided ideal of A. Then I is stable by involution and A/I is a C^*-algebra with the standard involution and the quotient norm.

PROOF. Let $(u_\alpha)_{\alpha \in \Lambda}$ be a family having the properties given in Lemma 6.4.4 and let $x \in I$. We have

$$\|u_\alpha x^* - x^*\| = \|x u_\alpha - x\|.$$

So $\lim_\alpha u_\alpha x^* = x^*$. But $u_\alpha x^* \in I$, and consequently $x^* \in I$. To prove that A/I is a C^*-algebra it is sufficient to prove that $|||\dot{x}|||^2 \leq |||\dot{x}^* \dot{x}|||$. We first prove that $|||\dot{x}||| = \lim_\alpha \|x - x u_\alpha\|$. If $y \in I$ we have $\lim_\alpha y u_\alpha = y$, so

$$\limsup_\alpha \|x - x u_\alpha\| = \limsup_\alpha \|x - x u_\alpha + y - y u_\alpha\| = \limsup_\alpha \|(x+y)(1-u_\alpha)\| \leq \|x+y\|,$$

because $\|1 - u_\alpha\| \leq 1$. Consequently

$$|||\dot{x}||| \geq \limsup_\alpha \|x - x u_\alpha\| \geq \liminf_\alpha \|x - x u_\alpha\| \geq \inf_{y \in I} \|x + y\| = |||\dot{x}|||.$$

So the assertion is proved. Now for every $z \in I$ we have

$$|||\dot{x}|||^2 = \lim_\alpha \|x - xu_\alpha\|^2 = \lim_\alpha \|(x - xu_\alpha)^*(x - xu_\alpha)\|$$

$$= \lim_\alpha \|x^*x - x^*xu_\alpha - u_\alpha x^*x + u_\alpha x^*xu_\alpha\|$$

$$= \lim_\alpha \|x^*x + z - zu_\alpha - x^*xu_\alpha - u_\alpha x^*x - u_\alpha z + u_\alpha zu_\alpha + u_\alpha x^*xu_\alpha\|$$

$$= \lim_\alpha \|(1 - u_\alpha)(x^*x + z)(1 - u_\alpha)\| \leq \|x^*x + z\|.$$

Hence $|||\dot{x}|||^2 \leq |||\dot{x}^*\dot{x}|||$. ◻

Consequently the Calkin algebra $\mathcal{L}(H)/\mathcal{LC}(H)$ is a C^*-algebra.

We finish with a small result related to Markov covariances.

THEOREM 6.4.6. Let a be in a C^*-algebra A. Then $\|e^{-ta}\| < 1$, for all $t \geq 0$, if and only if $a + a^* \geq 0$ and $\operatorname{Re} \operatorname{Sp} a > 0$.

PROOF. First we prove that the two conditions are necessary. Suppose $t \geq 0$. By Exercise III.15, we have

$$e^{-t(a+a^*)} = \lim_n (e^{-ta/n} e^{-ta^*/n})^n.$$

But $\|e^{-ta/n}\| = \|e^{-ta^*/n}\| < 1$, and consequently

$$\rho(e^{-t(a+a^*)}) = \|e^{-t(a+a^*)}\| \leq 1.$$

The spectrum of $a + a^*$ being real, we conclude that $a + a^* \geq 0$. Moreover $\rho(e^{-ta}) \leq \|e^{-ta}\| < 1$. So by the Holomorphic Functional Calculus, $\operatorname{Re} \operatorname{Sp} a > 0$. We now prove that the two conditions are sufficient. We have

$$1 - e^{-ta^*} e^{-ta} = \int_0^t e^{-sa^*}(a^* + a)e^{-sa} \, ds \geq 0 \qquad (1)$$

by hypothesis. So $\|e^{-ta}\|^2 = \|e^{-ta^*} e^{-ta}\| \leq 1$ and then $\Gamma(t) = \|e^{-ta}\| \leq 1$, for all $t \geq 0$. This implies, in particular, that $\Gamma(t)$ is decreasing on $[0, +\infty)$. In order to prove the result we have to prove that Γ cannot be equal to 1 on some interval $[0, u], u > 0$. Suppose $\Gamma(t) = 1$ for $0 \leq t \leq u$. By Theorem 6.2.17 there exists an extreme state σ on A such that $\sigma(e^{-ua^*} e^{-ua}) = 1$. Because of continuity and positivity of the extreme state, we conclude that $\sigma(e^{-sa^*}(a^* + a)e^{-sa}) = 0$ for $0 \leq s \leq u$, so $\sigma(e^{-ta^*} e^{-ta}) = 1$ for $0 \leq t \leq u$. Applying the Identity Principle to the entire function $\sigma(e^{-za^*}e^{-za})$, we conclude that $\sigma(e^{-ta^*}e^{-ta}) = 1$ for all $t \geq 0$. Hence

$$1 = \sigma(e^{-ta^*}e^{-ta}) \leq \|e^{-ta^*}e^{-ta}\| = \|e^{-ta}\|^2 \leq 1,$$

so $\|e^{-ta}\| = 1$, for all $t \geq 0$. Then by Theorem 3.2.8 we get

$$\rho(e^{-ta}) = \lim_{n \to \infty} \|e^{-tna}\|^{1/n} = 1,$$

a contradiction. The proof is now complete. $\square$

EXERCISE 1.   Prove that $A(\Delta)$ and $C(K)$ are not C^*-algebras.

EXERCISE 2.   Consider $M_n(\mathbf{C})$ with the involution m^* defined by the conjugate of the transpose of m, for $m \in M_n(\mathbf{C})$. Determine explicitly the C^*-algebra norm on $M_n(\mathbf{C})$.

EXERCISE 3.   Prove that a commutative C^*-algebra without unit is isometrically isomorphic to $C_0(\mathfrak{M})$.

EXERCISE 4.   Let A be a C^*-algebra without unit. Prove that there exists a unique C^*-algebra norm on the Banach algebra with unit $\tilde{A} = A \times \mathbf{C}$.

EXERCISE 5.   Prove that a C^*-algebra is semi-simple.

EXERCISE 6.   Let A, B be two C^*-algebras and let Φ be a $\star$-morphism from A onto B. Prove that Φ is isometric if and only if Φ is injective.

$\ast$EXERCISE 7.   Let A be a finite-dimensional C^*-algebra and let $x \in A$ be such that $\|x\| \leq 1$. Prove that there exist two unitary elements u_1, u_2 such that $x = \frac{u_1 + u_2}{2}$.

EXERCISE 8.   Let Γ be the unit circle and let u be the identity function $u(z) = z$ in $C(\Gamma)$. Prove that u is unitary but cannot be written as e^{ih}, for some continuous real function h defined on Γ.

$\ast\ast$EXERCISE 9.   Prove that in Theorem 6.2.13 and Corollary 6.2.14, U can be replaced by E. First use Theorem 6.2.13, then use the fact that

$$x = \frac{1}{2\pi i} \int_0^{2\pi} (\lambda 1 + x)(1 + \lambda x^*)^{-1}\, d\lambda$$

for x normal, $\|x\| < 1$, $|\lambda| = 1$, and that $(\lambda 1 + x)(1 + \lambda x^*)^{-1}$ is unitary with $\mathrm{Sp}(\lambda 1 + x)(1 + \lambda x^*)^{-1}$ not meeting the half-line $\{-\alpha\lambda : \alpha \geq 0\}$. If you do not succeed look at [1], pp. 109-113 where you will find some suggestions.

EXERCISE 10.   Let A be a C^*-algebra for a norm $\|\cdot\|$ and let $|\cdot|$ be another Banach algebra norm on A such that $|e^{ih}| = 1$, for all self-adjoint elements h. Prove that the two norms coincide. This result is improved upon in the following difficult exercise.

$\ast\ast\ast$EXERCISE 11.   Let A be a Banach algebra with involution such that $\|e^{ih}\| = 1$, for all self-adjoint elements h. Prove that A is a C^*-algebra for this norm. If you do not succeed look at [1], pp. 122-123.

EXERCISE 12. Let A be a separable C^*-algebra. Prove that there exists an isometric $\star$-morphism from A into $\mathfrak{L}(l^2)$.

EXERCISE 13. Let A be a C^*-algebra. Suppose that there exists on A another involution $x \to x^{\#}$ such that $\|xx^{\#}\| = \|x\|^2$, for all $x \in A$. Prove that $x^* = x^{\#}$, for all $x \in A$. Conclude that two C^*-algebras are isomorphic if and only if they are $\star$-isomorphic.

EXERCISE 14. Let A be a C^*-algebra and let $x \in A$. We define $V(x) = \{s(x) : x \in \mathfrak{S}\}$, where $\mathfrak{S}$ is the set of states. Prove that $V(x)$ is a compact and convex set containing the spectrum of x. If x is normal prove that $V(x) = \operatorname{co} \operatorname{Sp} x$. Give an example showing that in general $V(x) \neq \operatorname{co} \operatorname{Sp} x$.

EXERCISE 15. Let T be a quasi-nilpotent bounded operator on a Hilbert space H. Suppose that $\frac{T+T^*}{2}$ is compact. Prove that T is compact.

EXERCISE 16. Prove that a normal operator $T \in \mathfrak{L}(H)$ is compact if and only if it satisfies the following two conditions:

 (i) $\operatorname{Sp} T$ has at most 0 as a limit point,

 (ii) if $\lambda \neq 0$ then $\dim \mathcal{N}(T - \lambda I)$ is finite.

EXERCISE 17. Let $T \in \mathfrak{L}(H)$ be normal and compact. Prove that T has an eigenvalue λ with $|\lambda| = \|T\|$ and that $f(T)$ is compact if $f \in C(\operatorname{Sp} T)$ and $f(0) = 0$. To prove this last property use Mergelyan's theorem (see [7], p. 423).

EXERCISE 18. On l^2 prove that the right shift, with weight $\alpha_n = 1$, has no polar decomposition.

EXERCISE 19. Let A be a C^*-algebra and let B be a Banach algebra with involution. Suppose that Φ is an injective $\star$-morphism from A into B. Prove that $\|\Phi(x)\| \geq \|x\|$, for all $x \in A$. This result can be improved by the following exercise which is a generalization of Exercise IV.11.

***EXERCISE 20. Let A be a C^*-algebra and let B be a Banach algebra. Suppose that Φ is an injective morphism from A into B. Prove that there exists $C > 0$ such that $\|\Phi(x)\| \geq C\|x\|$, for all $x \in A$. This result is due to S.B. Cleveland. Using an idea of A. Rodríguez Palacios it can be proved with the help of Theorem 5.5.1.

Chapter VII

AN INTRODUCTION TO
ANALYTIC MULTIFUNCTIONS

As we explained in the Preface this chapter will be a quick incursion, as the crow flies, into the important new field of analytic multifunctions. A complete treatment would need a full book. In particular it would be necessary to recall the great number of results we need in the theory of functions of several complex variables. Of course we are not able to do this, in such a limited number of pages, so we shall suppose that the reader is familiar with all these prerequisites, which are contained for instance in [5,9] or in many other books on the field, and which are summarized in the Appendix, §2. In the last ten years the use of subharmonic functions in spectral theory and in the theory of uniform algebras has given a lot of interesting new results (see Chapters III and V for spectral theory; Exercises VII.8-9-10 and [10] for the theory of uniform algebras).

In the 1980s the important analytic tool of analytic multifunctions came back to life. The origin of this concept goes back to some work by K. Oka related to singularity sets in $\mathbf{C}^2$, and in some sense, has its roots in the work of Charles Puiseux, a student of Augustin Cauchy, who obtained, in relation with algebraic geometry, several interesting results about algebroid functions. Of course K. Oka never gave applications to functional analysis.

This concept was resuscitated in 1980 for the following reasons. In 1977 the author introduced the following conjecture: if f is an analytic function from a domain D of $\mathbf{C}$ into a Banach algebra and if the set of λ for which $\operatorname{Sp} f(\lambda)$ is at most countable has a positive capacity, then $\operatorname{Sp} f(\lambda)$ is at most countable for all λ in D. He also explained how this conjecture would solve another one (a generalization of a conjecture due to A. Pełczyński concerning C^*-algebras), which can be expressed in the following way: let A be a Banach algebra with involution, and suppose that

Sp h is at most countable for all $h = h^*$, then Sp x is at most countable for all x in A, and A has a "particular" algebraic structure.

Another reason for this revival is the discovery of the strange parallelism between Theorem 3.4.25 and the famous E. Bishop's theorem on analytic structure ([10], Chapter 11). In 1977, John Wermer and the author were able to extend Bishop's theorem using the subharmonic technique exploited in the proof of Theorem 3.4.25 (see *Capacity and Uniform Algebras*, Journal of Functional Analysis **28** (1978), pp. 386–400).

So, behind the resemblance of spectral theory and theory of fibres on a uniform algebra, new objects to discover were hidden: the analytic multifunctions.

§1. Definitions and General Properties

The convenient definition was discovered by Z. Słodkowski (*Analytic Set-Valued Functions and Spectra*, Mathematische Annalen **256** (1981), pp. 363–386). In fact, in his paper, Z. Słodkowski gave two definitions, the first one being valid only for analytic multifunctions from $\mathbf{C}$ into $\mathbf{C}$, the second one for analytic multifunctions from $\mathbf{C}^n$ into $\mathbf{C}^m$, and he proved their equivalence for $m = n = 1$. We shall use the second definition as it leads to interesting results very quickly.

The majority of results and ideas mentioned in this section are due to B. Aupetit, T. J. Ransford, Z. Słodkowski, H. Yamaguchi and A. Zraïbi.

DEFINITION. Let K be a mapping from an open subset U of $\mathbf{C}^n$ into the set of non-empty compact subsets of $\mathbf{C}^m$. We shall say that K is an *analytic multifunction* on U if it satisfies the following properties:

(i) K is upper semicontinuous on U,

(ii) for every relatively compact open subset V of U and every function ϕ plurisubharmonic on a neighbourhood of the restriction of the graph of K to V, the function

$$\psi(\lambda) = \max\{\phi(z) : z \in K(\lambda)\}$$

is plurisubharmonic on V.

A function is plurisubharmonic if and only if it is locally plurisubharmonic, and consequently a multifunction is analytic if and only if it is locally analytic. In (ii) we may suppose that ϕ is a C^∞-strictly plurisubharmonic function since ϕ can be approximated by such functions.

EXAMPLES.

(1) If h is holomorphic on $D \subset \mathbf{C}^n$, with values in $\mathbf{C}^n$, then $\lambda \mapsto \{h(\lambda)\}$ is an analytic multifunction on D.

(2) Let K_0, K_1 be two compact subsets of $\mathbf{C}^n$. For $\lambda \in D \subset \mathbf{C}$, the multifunction $K(\lambda) = \lambda K_0 + K_1$ is analytic on D.

(3) Let f be an analytic function from $D \subset \mathbf{C}$ into $M_n(\mathbf{C})$. Then $\lambda \mapsto \operatorname{Sp} f(\lambda)$ is an analytic multifunction on D (see Exercise VII.2).

Much more important results will be given later.

An interesting class of multifunctions is the following.

DEFINITION. An upper semicontinuous multifunction K from $D \subset \mathbf{C}^n$ into $\mathbf{C}^m$ is said to have *holomorphic selections* if, for each $\lambda_0 \in D$ and each $z_0 \in \partial K(\lambda_0)$, there exists h holomorphic on a neigbourhood U of λ_0, with values in $\mathbf{C}^m$, such that $z_0 = h(\lambda_0)$ and $h(\lambda) \in K(\lambda)$, for $\lambda \in U$.

Such multifunctions having holomorphic selections are continuous analytic multifunctions (see Exercise VII.3). Examples (1) and (2) are in this class, but not (3) globally on D because of the branching points. The following theorem has a very technical proof which will not be given here. We refer the reader to the papers of B. Aupetit, B.Aupetit & A.Zraïbi, Z. Słodkowski, and T.J. Ransford.

THEOREM 7.1.1. *The following properties hold.*

(i) *If (K_p) is a sequence of analytic multifunctions defined on $D \subset \mathbf{C}^n$ with values in $\mathbf{C}^m$, such that $K_{p+1}(\lambda) \subset K_p(\lambda)$ for each $\lambda \in D$, then $K = \cap_{p=1}^{\infty} K_p$ is an analytic multifunction from D into $\mathbf{C}^m$.*

(ii) *If $K_1, \ldots, K_p$ are analytic multifunctions from $D \subset \mathbf{C}^n$ into $\mathbf{C}^m$ then $K = K_1 \cup \ldots \cup K_p$ is an analytic multifunction from D into $\mathbf{C}^m$.*

(iii) *If K is an analytic multifunction from $D \subset \mathbf{C}^n$ into $\mathbf{C}^m$ and if L is an analytic multifunction from $G \subset \mathbf{C}^m$ into $\mathbf{C}^k$, where G is an open set containing all the $K(\lambda)$ for $\lambda \in D$, then $L \circ K$ defined by*

$$(L \circ K)(\lambda) = \{L(z) : z \in K(\lambda)\}$$

is an analytic multifunction from D into $\mathbf{C}^k$.

(iv) *If $K_1, \ldots, K_p$ are analytic multifunctions from $D \subset \mathbf{C}^n$ into $\mathbf{C}^m$, then $K = K_1 \times \cdots \times K_p$ is an analytic multifunction from D into $\mathbf{C}^{mp}$.*

(v) *Let K be an upper semicontinuous multifunction from $D \subset \mathbf{C}^n$ into $\mathbf{C}^m$. Then K is an analytic multifunction on D if and only if $t \mapsto K(at + b)$ is an analytic multifunction on $\{t : t \in \mathbf{C}, at + b \in D\}$, for every $a, b \in \mathbf{C}^n$.*

THEOREM 7.1.2. *Let K be an analytic multifunction from $D \subset \mathbf{C}^n$ into $\mathbf{C}^m$ and suppose that L is an upper semicontinuous multifunction from D into $\mathbf{C}^m$ such that $\partial L(\lambda) \subset K(\lambda) \subset L(\lambda)$, for each $\lambda \in D$. Then L is an analytic multifunction. In particular $K\hat{\ }$ is an analytic multifunction.*

PROOF. Let D_1 be open in D and let ϕ be plurisubharmonic on a neighbourhood of the graph of L restricted to D_1. If $\lambda \in D_1$ then we have, by the Maximum Principle,

$$\psi(\lambda) = \max\{\phi(\lambda, z) \colon z \in L(\lambda)\} = \max\{\phi(\lambda, z) \colon z \in \partial L(\lambda)\}$$
$$\leq \max\{\phi(\lambda, z) \colon z \in K(\lambda)\} = \psi_1(\lambda) \leq \max\{\phi(\lambda, z) \colon z \in L(\lambda)\}.$$

So $\psi(\lambda) = \psi_1(\lambda)$, and ψ_1 is subharmonic by hypothesis. Obviously $K\hat{\ }$ satisfies the two inclusions, and it is easy to verify that it is upper semicontinuous. □

THEOREM 7.1.3. *Let K be an analytic multifunction from $D \subset \mathbf{C}$ into $\mathbf{C}$. Then $\lambda \mapsto \log \rho(K(\lambda))$, $\lambda \mapsto \log \delta_n(K(\lambda))$ for $n \geq 1$, and $\lambda \mapsto \log c(K(\lambda))$ are subharmonic on D (where c denotes the capacity).*

PROOF. For the first function, this is immediate from the definition using the plurisubharmonic function $\phi(\lambda, z) = \log |z|$. For the second one, let

$$\phi(\lambda, z_1, \ldots, z_{n+1}) = \frac{2}{n(n+1)} \sum_{1 \leq i < j \leq n+1} \log |z_i - z_j|,$$

which is plurisubharmonic on $\mathbf{C}^{n+2}$. By Theorem 7.1.1 (iv), we know that

$$\lambda \mapsto K(\lambda) \times \ldots \times K(\lambda), \ n+1 \text{ times},$$

is an analytic multifunction. So, by definition,

$$\log \delta_n(K(\lambda)) = \max\{\phi(\lambda, z_1, \ldots, z_{n+1}) : z_1 \in K(\lambda), \ldots, z_{n+1} \in K(\lambda)\}$$

is plurisubharmonic. By Theorem A.1.22, we know that $c(K(\lambda))$ is the decreasing limit of $\delta_n(K(\lambda))$, when n goes to infinity. So, by Theorem A.1.1 (vi) for subharmonic functions, we get the last result. □

Let D be an open subset of $\mathbf{C}^n$, h be holomorphic on D and $\Omega = \{(\lambda, z) \colon \lambda \in D, z \neq h(\lambda)\}$ be the complement of the graph of h. Considering the function $g(\lambda, z) = 1/(z - h(\lambda))$, which is holomorphic on Ω, it is obvious that Ω is an open set of holomorphy and so it is pseudoconvex. F. Hartogs proved the converse, namely that if h is a locally bounded function from D into $\mathbf{C}$ such that Ω is an open set of holomorphy, then h is holomorphic on D. The original proof uses subharmonic arguments (see R. Narasimhan, *Several Complex Variables*, p. 56). With Theorem 7.1.10, Hartogs's result can be reduced to the following generalization.

THEOREM 7.1.4. *Let h be a function from $D \subset \mathbf{C}^n$ into $\mathbf{C}^m$. Then h is holomorphic on D if and only if $\lambda \mapsto \{h(\lambda)\}$ is an analytic multifunction on D.*

PROOF. It is obvious that h holomorphic implies that $\lambda \mapsto \{h(\lambda)\}$ is an analytic multifunction. Suppose now that $\lambda \mapsto \{h(\lambda)\}$ is an analytic multifunction on D. By Theorem 7.3.1 (v) and F. Hartogs's theorem on functions which are holomorphic in each variable ([5], p. 28 or Theorem A.2.1), it is sufficient to prove the theorem for $m = n = 1$. Because the result is purely local, we suppose that we are on a closed disk $\overline{\Delta}$ included in D and we prove that h is holomorphic on Δ. Let A be the set of $f \in C(\overline{\Delta})$ such that there exists an analytic multifunction K defined on a neighbourhood of $\overline{\Delta}$ with $K(\lambda) = \{f(\lambda)\}$, for all $\lambda \in \overline{\Delta}$. By Theorem 7.1.1, it follows that A is a subalgebra of $C(\overline{\Delta})$ containing the polynomials. By the Maximum Principle we have, for $f \in A$,

$$\max\{|f(\lambda)|: \lambda \in \overline{\Delta}\} = \max\{|f(\lambda)|: \lambda \in \partial\Delta\}.$$

So, by W. Rudin's characterization of holomorphic functions ([7], p. 280), all the elements of A, and in particular h, are holomorphic on Δ. $\blacksquare$

A great number of results obtained in Chapter III for $\mathrm{Sp}\, f(\lambda)$, where f is an analytic function from a domain $D \subset \mathbf{C}$ into a Banach algebra, can be extended with very similar arguments to analytic multifunctions (see Exercise VII.6). The three following theorems are very important.

THEOREM 7.1.5 (LOCALIZATION PRINCIPLE). *Let D be a domain of $\mathbf{C}^n$ and let K be an analytic multifunction from D into $\mathbf{C}^m$. Suppose that there exist two disjoint open sets U, V in $\mathbf{C}^m$ such that $K(\lambda) \subset U \cup V$, for every $\lambda \in D$. Then either $K(\lambda) \cap U = \emptyset$, for every $\lambda \in D$, or $K(\lambda) \cap U \neq \emptyset$, for every $\lambda \in D$. In the latter case $L(\lambda) = K(\lambda) \cap U$ defines an analytic multifunction on D.*

PROOF. The two open sets $D \times U$ and $D \times V$ are disjoint so ϕ_0, defined to be 1 on the former one and 0 on the latter, is plurisubharmonic on a neighbourhood of the graph of K. Let $\psi_0(\lambda) = \max\{\phi_0(\lambda, z): z \in K(\lambda)\}$. Then ψ_0 is plurisubharmonic on D. Also $\psi_0(\lambda) = 0$ if $K(\lambda) \cap U = \emptyset$, and $\psi_0(\lambda) = 1$ if $K(\lambda) \cap U \neq \emptyset$. Then ψ_0 is either identically zero or identically one on D, so we obtain the first part of the theorem. Now let D_1 be an open subset of D and ϕ be plurisubharmonic on a neighbourhood of the graph of L restricted to D_1. Without loss of generality we may suppose that ϕ is defined on an open subset of $D \times U$. So we may extend ϕ to a function ϕ_1 plurisubharmonic on a neighbourhood of the graph of K such that $\phi_1(\lambda) = \phi(\lambda)$ on the open subset of $D \times U$ and $\phi_1(\lambda) = -\infty$ on $D \times V$. Then $\psi(\lambda) = \max\{\phi(\lambda, z): z \in L(\lambda)\} = \max\{\phi_1(\lambda, z): z \in K(\lambda)\}$. So ψ is plurisubharmonic on D_1, and hence L is analytic multivalued. $\blacksquare$

THEOREM 7.1.6 (HOLOMORPHIC VARIATION OF ISOLATED POINTS). *Let D be a domain of $\mathbf{C}^n$ and K be an analytic multifunction from D into $\mathbf{C}^m$. Suppose that U is an open subset of $\mathbf{C}^m$ such that $\#(K(\lambda) \cap U) = 1$ and $K(\lambda) \cap \partial U = \emptyset$, for $\lambda \in D$. Then there exists h holomorphic on D, with values in $\mathbf{C}^m$, such that $K(\lambda) \cap U = \{h(\lambda)\}$.*

PROOF. We apply Theorem 7.1.5 and Theorem 7.1.4. ◻

THEOREM 7.1.7 (SCARCITY OF ELEMENTS WITH FINITE VALUES). *Let D be a domain of $\mathbf{C}^n$ and K be an analytic multifunction from D into $\mathbf{C}^m$. Then either $\{\lambda : \lambda \in D, \#K(\lambda) < \infty\}$ is pluripolar in D or there exist an integer N and a closed analytic subvariety F of D such that $\#K(\lambda) = N$ on $D \setminus F$ and $\#K(\lambda) < N$ on F. Moreover, in this last situation, for each $\lambda_0 \in D \setminus F$ there exist N functions $h_1, \ldots, h_N$ with values in $\mathbf{C}^m$, holomorphic on a neighbourhood of λ_0, such that $K(\lambda) = \{h_1(\lambda), \ldots, h_N(\lambda)\}$ on this neighbourhood.*

PROOF. The proof is very similar, except for some technical points, to the proof of Theorem 3.4.25. We leave it to the reader as an exercise. ◻

In order to prove Z. Słodkowski's result we need two theorems, the first one being an improvement of the result proved in Exercise VII.1, and the second one resulting from the classical theorem on the exhaustion of pseudoconvex open sets by regular strictly pseudoconvex open sets (Theorem A.2.16). The following definition extends the notion of holomorphic selections. For instance, example (3) has no holomorphic selections at the branching points but has good selections everywhere, where a good selection is defined as follows.

DEFINITION. An upper semicontinuous multifunction K from $D \subset \mathbf{C}^n$ into $\mathbf{C}$ is said to have *good selections* if, for each $\lambda_0 \in D$ and each $z_0 \in \partial K(\lambda_0)$, either there exist $r > 0$ and h holomorphic for $|\lambda - \lambda_0| < r$, with complex values, such that $h(\lambda_0) = z_0$ and $h(\lambda) \in K(\lambda)$ for $|\lambda - \lambda_0| < r$, or there exist $s > 0$ and k holomorphic for $|z - z_0| < s$, with values in D, such that $k(z_0) = \lambda_0$ and $z \in K(k(z))$ for $|z - z_0| < s$.

THEOREM 7.1.8. *Let K be an upper semicontinuous multifunction from $D \subset \mathbf{C}$ into $\mathbf{C}$, having good selections. Then K is a continuous analytic multifunction on D.*

PROOF. We leave the proof of continuity to the reader. Let ϕ be plurisubharmonic on a neighbourhood of the graph of K on D, and let $\psi(\lambda) = \max\{\phi(\lambda, z) : z \in K(\lambda)\}$.

Then ψ is upper semicontinuous. Let $\overline{\Delta}$ be a closed disk included in D and let p be a polynomial such that $\psi(\lambda) \leq \mathrm{Re}\, p(\lambda)$ on $\partial\Delta$. In order to show that ψ is subharmonic we have to prove that the previous inequality is true on Δ. Suppose that there exists $\lambda \in \Delta$ such that $\psi(\lambda) > \mathrm{Re}\, p(\lambda)$, and let λ_0 in Δ be such that $\psi(\lambda_0) - \mathrm{Re}\, p(\lambda_0) = \sup\{\psi(\lambda) - \mathrm{Re}\, p(\lambda) : \lambda \in \overline{\Delta}\}$. Then $E = \{\lambda : \lambda \in \overline{\Delta}, \psi(\lambda_0) - \mathrm{Re}\, p(\lambda_0) = \psi(\lambda) - \mathrm{Re}\, p(\lambda)\}$ is compact in Δ, so we can suppose that $\mathrm{dist}(E, \partial\Delta) = \mathrm{dist}(\lambda_0, \partial\Delta) = 2\epsilon$. We have $\psi(\lambda_0) = \phi(\lambda_0, z_0)$ for some $z_0 \in \partial K(\lambda_0)$. Because K has good selections, in the first situation there exist r such that $0 < r < \epsilon$ and h holomorphic for $|\lambda - \lambda_0| < r$ such that $h(\lambda_0) = z_0$ and $h(\lambda) \in K(\lambda)$. Then $u(\lambda) = \phi(\lambda, h(\lambda)) - \mathrm{Re}\, p(\lambda) - \psi(\lambda_0) + \mathrm{Re}\, p(\lambda_0)$ is subharmonic for $|\lambda - \lambda_0| < r$, and moreover $u(\lambda_0) = 0$ and $u(\lambda) \leq 0$ for $|\lambda - \lambda_0| < r$. By the Maximum Principle, $u(\lambda) = 0$ for $|\lambda - \lambda_0| < r$, but this contradicts the fact that $\mathrm{dist}(E, \partial\Delta) = \mathrm{dist}(\lambda_0, \partial\Delta)$. In the second situation there exist $s > 0$ and k holomorphic for $|z - z_0| < s$ such that $k(B(z_0, s)) \subset B(\lambda_0, s)$, $k(z_0) = \lambda_0$ and $z \in K(k(z))$ for $|z - z_0| < s$. Then $v(z) = \phi(k(z), z) - \mathrm{Re}\, p(k(z)) - \psi(\lambda_0) + \mathrm{Re}\, p(\lambda_0)$ is subharmonic for $|z - z_0| < r$, satisfies $v(z_0) = 0$, and $v(s) \leq 0$ for $|z - z_0| < s$. By the Maximum Principle, $v(z) = 0$ for $|z - z_0| < s$. This is a contradiction, as k is open, then $k(B(z_0, s))$ contains a disk $B(\lambda_0, \alpha)$ with $\alpha > 0$ and so on $B(\lambda_0, \alpha)$ there exists some $k(z)$ with $v(z) < 0$. $\square$

THEOREM 7.1.9. *Let K be an upper semicontinuous multifunction from $D \subset \mathbf{C}$ into $\mathbf{C}$ such that the complement of its graph is pseudoconvex in $\mathbf{C}^2$. Then there exist an increasing sequence of relatively compact open subsets D_n of D, exhausting D, and a sequence (K_n) of analytic multifunctions, respectively defined on D_n, having good selections, and such that $K_{n+1}(\lambda) \subset K_n(\lambda)$ for each $\lambda \in D_n$ and $K(\lambda) = \lim_{n\to\infty} K_n(\lambda)$ for each $\lambda \in D$. In particular K is an analytic multifunction on D.*

PROOF. Let $D_n = \{\lambda : \lambda \in D, \mathrm{dist}(\lambda, \partial D) > 1/n\}$. Then $\overline{D}_n$ is compact in D and $D = \cup_{n=1}^{\infty} D_n$. The set $C_n = \cup_{\lambda \in \overline{D}_n} K(\lambda)$ is compact, so we can construct an increasing sequence of open disks U_n such that $U_n \supset C_n$. Let $\Omega = \{(\lambda, z) : \lambda \in D, z \notin K(\lambda)\}$. Because Ω is pseudoconvex, there exists $\phi \in C^{\infty}(\Omega)$, ϕ strictly plurisubharmonic on Ω, such that $\Omega_k = \{(\lambda, z) : (\lambda, z) \in \Omega, \phi(\lambda, z) < k\}$ is relatively compact and satisfies $\Omega = \cup_{k=1}^{\infty} \Omega_k$. So we can construct inductively an increasing sequence (Ω_{k_n}) such that $\Omega_{k_1} \supset \overline{D}_1 \times \partial U_1, \ldots, \Omega_{k_{n+1}} \supset (\overline{D}_{n+1} \times \partial U_{n+1}) \cup (\overline{D}_n \times (\overline{U}_{n+1} \setminus U_n)), \ldots$, for all integers n. Then we define $K_n(\lambda)$ on D_n by $K_n(\lambda) = \{z : z \in U_n, (\lambda, z) \notin \Omega_{k_n}\}$. This set $K_n(\lambda)$ is compact and non-empty because $K(\lambda) \subset K_n(\lambda)$ for all $\lambda \in D_n$. By Theorem A.2.17, K_n has good selections so, by Theorem 7.1.8, K_n is analytic multivalued on D_n. Let $\lambda \in D_n$ and $z \in K_{n+1}(\lambda)$, so that $z \in U_{n+1}$ and $(\lambda, z) \notin \Omega_{k_{n+1}}$. If $z \notin U_n$ then $(\lambda, z) \in D_n \times (U_{n+1} \setminus U_n) \subset \Omega_{k_{n+1}}$, which gives a contradiction. So $z \in U_n$ and $(\lambda, z) \notin \Omega_{k_n}$, hence $z \in K_n(\lambda)$. Because $\cup_{n=1}^{\infty} \Omega_{k_n} = \Omega$, it is obvious that for $\lambda \in D_n$ we have $\cap_{m \geq n} K_n(\lambda) = K(\lambda)$, and hence $K = \lim_{m \to \infty} K_n$. By Theorem 7.1.1 (i), K is analytic multivalued on D. $\square$

THEOREM 7.1.10 (Z.SŁODKOWSKI). *Let K be an upper semicontinuous multifunction from $D \subset \mathbb{C}$ into $\mathbb{C}$. Then the following are equivalent:*

(i) K is an analytic multifunction,

(ii) $-\log \text{dist}(z, K(\lambda))$ is plurisubharmonic on Ω, the complement of its graph,

(iii) Ω is pseudoconvex in $\mathbb{C}^2$.

PROOF. By Exercise VII.5, (i) implies (ii). If (ii) is verified then $-\log \text{dist}(z, K(\lambda))$ is a plurisubharmonic vertical function for Ω, so Ω is pseudoconvex (see Theorem A.2.14). If (iii) is verified then by Theorem 7.1.9, K is analytic multivalued. $\square$

The following result is in fact equivalent to a classical result due to F. Hartogs. It gives interesting examples of discontinuous analytic multifunctions.

THEOREM 7.1.11. *Let D be open in $\mathbb{C}$ and ϕ be defined on D with values in $\mathbb{R} \cup \{-\infty\}$. Then the multifunction K, defined by $K(\lambda) = \{z : |z| \leq e^{\phi(\lambda)}\}$ on D, is an analytic multifunction on D if and only if ϕ is subharmonic on D.*

PROOF. If K is an analytic multifunction on D then, by Theorem 7.1.3, $\log \rho(K(\lambda)) = \phi(\lambda)$ is subharmonic on D. Conversely if ϕ is subharmonic, then ϕ is upper semicontinuous, so K is upper semicontinuous on D. By Theorem 7.1.10 we have to show that

$$\Omega = \{(\lambda, z) : \lambda \in D, |z| > e^{\phi(\lambda)}\}$$

is pseudoconvex. We know that $\{(\lambda, z) : \lambda \in D, z \neq 0\}$ is pseudoconvex. The function $(\lambda, z) \mapsto |z|e^{\phi(\lambda)}$ is plurisubharmonic on $D \times \mathbb{C}$, which is pseudoconvex, and so the set $\{(\lambda, z) : \lambda \in D, |z|e^{\phi(\lambda)} < 1\}$ is pseudoconvex. Then $\Omega' = \{(\lambda, z) : \lambda \in D, z \neq 0, |z|e^{\phi(\lambda)} < 1\}$ is pseudoconvex, being the intersection of two pseudoconvex sets. But $\lambda' = \lambda$, $z' = 1/z$ defines a biholomorphic transformation from Ω' onto Ω. So Ω is pseudoconvex. $\square$

COROLLARY 7.1.12 (F. HARTOGS). *Let D be open in $\mathbb{C}$ and ψ be a positive function on D such that $\{(\lambda, z) : \lambda \in D, |z| < \psi(\lambda)\}$ is pseudoconvex in $\mathbb{C}^2$. Then $-\log \psi(\lambda)$ is subharmonic on D.*

Taking $\phi(\lambda) = \sum_{n=1}^{\infty} \dfrac{\log|\lambda - 1/n| + \log n}{n^3}$, which is well-defined and subharmonic on $\mathbf{C}$, we have $\phi(1/n) = -\infty$ for $n = 1, 2, \ldots$, and $\phi(0) = 0$. The corresponding analytic multivalued function K satisfies $K(1/n) = \{0\}$ and $K(0) = \{z : |z| \leq 1\}$, so it is discontinuous at 0. By condensation of the singularities it is even possible to construct an analytic multifunction K discontinuous on a dense set.

We now finish this section with the most important theorems: these are the motivation behind the theory of analytic multifunctions.

THEOREM 7.1.13 (B.AUPETIT-Z.SLODKOWSKI). *Let f be an analytic function from an open set $D \subset \mathbf{C}$ into a Banach algebra. Then $\lambda \mapsto \operatorname{Sp} f(\lambda)$ is an analytic multifunction on D.*

PROOF. By Theorem 3.4.2, $\lambda \mapsto \operatorname{Sp} f(\lambda)$ is upper semicontinuous on D. The function $(\lambda, z) \mapsto (z1 - f(\lambda))^{-1}$ is analytic on Ω, so $\phi(\lambda, z) = \|(z1 - f(\lambda))^{-1}\| - \log \operatorname{dist}(\lambda, \partial D)$ is plurisubharmonic on Ω. We now show that $\lim \phi(\lambda, z) = +\infty$ when (λ, z) goes to $(\lambda_0, z_0) \in \partial\Omega$. Suppose that there exist $M > 0$ and two sequences (λ_n), (z_n) such that $\lim \lambda_n = \lambda_0$, $\lim z_n = z_0$, $(\lambda_n, z_n) \in \Omega$ and $\phi(\lambda_n, z_n) \leq M$. Because $\operatorname{dist}(\lambda_n, \partial D) \geq e^{-M}$ we conclude that $\lambda_0 \in D$, so $z_0 1 - f(\lambda_0)$ is not invertible. But we have

$$z_0 1 - f(\lambda_0) = z_0 1 - z_n 1 + z_n 1 - f(\lambda_n) + f(\lambda_n) - f(\lambda_0)$$
$$= (z_n 1 - f(\lambda_n))[1 + (z_n 1 - f(\lambda_n))^{-1} u(\lambda_n, z_n)].$$

If λ_n goes to λ_0 and z_n goes to z_0, then $u(\lambda_n, z_n)$ goes to 0 and we have

$$\|(z_n 1 - f(\lambda_n))^{-1} u(\lambda_n, z_n)\| \leq [M + \log \operatorname{dist}(\lambda_n, \partial D)] u(\lambda_n, z_n) < 1,$$

for n large enough. Thus $z_0 1 - f(\lambda_0)$ is invertible, which is a contradiction. By Theorem A.2.11, Ω is pseudoconvex, so $\lambda \mapsto \operatorname{Sp} f(\lambda)$ is analytic multivalued on D, by Theorem 7.1.10. $\square$

We recall that the Šilov boundary is defined in Exercice IV.10.

THEOREM 7.1.14 (Z.SLODKOWSKI). *Let A be a commutative Banach algebra. Denote by $\mathfrak{M}$ its set of characters and by $\check{S}$ its Šilov boundary. Let f, g be two elements of A and let W be a component of $\hat{f}(\mathfrak{M}) \setminus \hat{f}(\check{S})$. Then $\lambda \mapsto g(f^{-1}(\lambda)) = \{\chi(g) : \chi(f) = \lambda\}$ is an analytic multifunction on W.*

The proof is too complicated to be given here. It can be found in several papers by Z. Slodkowski.

A weak form of this theorem, due to J. Wermer, will be found in Exercise VII.8.

Theorem 7.1.14 and Theorem 7.1.7 imply immediately a strong generalization of E. Bishop's theorem on analytic structure ([10], Chapter 11).

Finally, why are Theorems 7.1.10, 7.1.13, 7.1.14 of interest? At first glance it seems very strange to start with some spectral problems or some questions relating to uniform algebras and then to consider the rather difficult theory of pseudoconvex open sets. What do we gain by doing this? The main reason is that many problems in spectral theory and in the theory of uniform algebras are reduced to purely geometrical problems concerning pseudoconvex open subsets of $\mathbf{C}^2$. Theorem 7.1.6 is a good illustration of this. In the next two sections we shall discover new ones.

§2. The Oka-Nishino Theorem and Its Applications

We now intend to give the proof of the very important Oka-Nishino theorem. The original proof due to T. Nishino, contains many obscure points. The following presentation is largely inspired by the ideas of B. Aupetit, T. Nishino, T.J. Ransford, Z. Słodkowski and J. Zemánek. Very often it will be rather superficial since the arguments involved are extremely technical.

DEFINITION. Let K be an analytic multifunction from $D \subset \mathbf{C}$ into $\mathbf{C}$. We say that $z_0 \in K(\lambda_0)$ is a *good isolated point* of $K(\lambda_0)$ if there exist $r, s > 0$ such that $\overline{B}(z_0, s) \cap K(\lambda_0) = \{z_0\}$, and $B(z_0, s) \cap K(\lambda)$ is finite for $|\lambda - \lambda_0| < r$.

This definition implies that z_0 is isolated in $K(\lambda_0)$. But conversely an isolated point of $K(\lambda_0)$ is not necessarily a good isolated point. To see this, consider the following example of the analytic multifunction defined for $\lambda \in \mathbf{C}$ by

$$K(\lambda) = \{0\} \cup \left\{ z : \frac{|\lambda|}{2} \le |z| \le |\lambda| \right\}.$$

Of course 0 is isolated in $K(0)$, but it is not a good isolated point. In fact this definition of a good isolated point is purely geometrical. It means that in a neighbourhood of (λ_0, z_0), the graph of K is an analytic variety. This notion is closely related to that of extension points for pseudoconvex sets which we now define.

DEFINITION. Let Ω be a pseudoconvex open subset of $\mathbf{C}^2$. We say that a point $a \in \mathbf{C}^2 \setminus \Omega$ is an *extension point* for Ω if there exist an open neighbourhood U of a and a non-zero holomorphic function f on U, with values in $\mathbf{C}$, such that

$$U \setminus \Omega = \{z : z \in U, f(z) = 0\}.$$

We define Ω' to be the union of Ω and its extension points. Clearly Ω' is open and contains Ω.

THEOREM 7.2.1.   *Let K be an analytic multifunction from $D \subset \mathbf{C}$ into $\mathbf{C}$. If z_0 is a good isolated point of $K(\lambda_0)$ then (λ_0, z_0) is an extension point for $\Omega = \{(\lambda, z) : \lambda \in D, z \notin K(\lambda)\}$. Conversely, if $z_0 \in \partial K(\lambda_0)$ and (λ_0, z_0) is an extension point for Ω, then z_0 is a good isolated point of $K(\lambda_0)$.*

PROOF.   Suppose first that z_0 is a good isolated point of $K(\lambda_0)$. By Theorem 7.1.5 and Theorem 7.1.7 applied to $\lambda \mapsto B(z_0, s) \cap K(\lambda)$ for $|\lambda - \lambda_0| < r$, we conclude that there exist a smallest integer n and a closed discrete set E in $B(\lambda_0, r)$ such that $\#(B(z_0, s) \cap K(\lambda)) = n$ for $\lambda \in B(\lambda_0, r) \setminus E$ and $\#(B(z_0, s) \cap K(\lambda)) < n$ for $\lambda \in E$. Taking a smaller r if necessary, we may suppose that $E \cap B(\lambda_0, r) \subset \{\lambda_0\}$. So $B(z_0, s) \cap K(\lambda) = \{h_1(\lambda), \dots, h_n(\lambda)\}$, where $h_1, \dots, h_n$ are holomorphic for $0 < |\lambda - \lambda_0| < r$. The symmetric functions in $h_1(\lambda), \dots, h_n(\lambda)$ define global holomorphic functions for $0 < |\lambda - \lambda_0| < r$. The singularity at λ_0 is removable for these symmetric functions because they can be extended continuously, using a similar argument to that in the proof of theorem 3.4.5. So $f(\lambda, z) = \prod_{i=1}^{n}(z - h_i(\lambda))$, which is expressed only with the symmetric functions of the $h_i(\lambda)$, is holomorphic on $B(\lambda_0, r) \times B(z_0, s)$. But $z \in K(\lambda)$, for $|\lambda - \lambda_0| < r$, $|z - z_0| < s$, is equivalent to saying that $f(\lambda, z) = 0$. So (λ_0, z_0) is an extension point for Ω.

Suppose now that $z_0 \in \partial K(\lambda_0)$ and that (λ_0, z_0) is an extension point for Ω. Let f and U be as in the definition of an extension point. If $h(z) = f(\lambda_0, z) \equiv 0$, locally around z_0, then z_0 is interior to $K(\lambda_0)$, which is a contradiction. So the zeros of h are isolated. Let $s > 0$ be such that $\overline{B}(z_0, s) \cap K(\lambda_0) = \{z_0\}$ and let $r > 0$ be such that $|\lambda - \lambda_0| < r$ implies $\partial B(z_0, s) \cap K(\lambda) = \emptyset$. We can also suppose that r is small enough for the zeros of $f(\lambda, z)$ to be isolated for all λ fixed satisfying $|\lambda - \lambda_0| < r$. Consequently $|\lambda - \lambda_0| < r$ implies that $B(z_0, s) \cap K(\lambda)$ is finite. ∎

LEMMA 7.2.2.   *Let M, N be open disks in $\mathbf{C}$ and $U = N \times M$. Let f be holomorphic on U, with values in $\mathbf{C}$, not identically zero on U. Let Z be the set of $(z, \eta) \in \mathbf{C}^2$ such that*

(i) $D(z, \eta) = \{w : w \in M, z + \eta w \in N\} \neq \emptyset$,

(ii) $g(w) = f(z + \eta w, w)$ *is identically zero on $D(z, \eta)$.*

Then Z is at most countable.

PROOF.   Suppose Z to be infinite and uncountable. Then there exists $(z_0, \eta_0) \in Z$ such that each of its neighbourhoods contains an infinite and uncountable subset of Z. Let $D = D(z_0, \eta_0)$, which is open in $\mathbf{C}$. Since f is not identically zero on U, the set

$$W = \{w : w \in M, g(\zeta) = f(\zeta, w) \equiv 0 \text{ on } N\}$$

is closed and discrete. So $D \setminus W$ is open and non-empty. Let $w_0 \in D \setminus W$ and $r > 0$ be such that

$$|w - w_0| < r \text{ and } |z - z_0| < r \text{ and } |\eta - \eta_0| < r \text{ imply } w \in D \setminus W \text{ and } z + \eta w \in N.$$

By hypothesis there are uncountably many pairs $(z, \eta) \in Z$ such that $|z - z_0| < r$ and $|\eta - \eta_0| < r$. But for each w with $|w - w_0| < r$, the set

$$\{z + \eta w : (z, \eta) \in Z, \ |z - z_0| < r, \ |\eta - \eta_0| < r\}$$

is contained in the zero set of $g(\zeta) = f(\zeta, w)$, so is an at most countable subset of N. Thus we get a contradiction. $\square$

THEOREM 7.2.3 (K.OKA-T.NISHINO). *Let Ω be a pseudoconvex open subset of $\mathbb{C}^2$. Then Ω' is also pseudoconvex.*

SKETCH OF PROOF. Suppose that Ω' is not pseudoconvex. Then we may choose $r > 1$ and $F : \mathbb{C} \times B(0, r) \mapsto \mathbb{C}^2$ such that:

(i) F is a biholomorphic mapping onto an an open subset of $\mathbb{C}^2$,

(ii) $|z| \leq 1$ and $|w| = 1$ implies $F(z, w) \in \Omega'$,

(iii) $|z| < 1$ and $|w| \leq 1$ implies $F(z, w) \in \Omega'$,

(iv) $F(1, 0) \notin \Omega'$,

(v) if $F = (F_1, F_2)$ then the two functions $f_i : B(0, r) \mapsto \mathbb{C}$, defined by $f_i(w) = F_i(1, w)$, are non-constant.

Let $\Lambda = F^{-1}(\Omega)$. Then we have Λ pseudoconvex in $\mathbb{C} \times B(0, r)$. Also, since F is biholomorphic, we have $\Lambda' = F^{-1}(\Omega')$. Thus

$$|z| < 1 \text{ and } |w| \leq 1 \text{ imply } (z, w) \in \Lambda',$$
$$|z| \leq 1 \text{ and } |w| = 1 \text{ imply } (z, w) \in \Lambda',$$
$$\text{and } (1, 0) \notin \Lambda'.$$

We consider the set $R = \{w : w \in \mathbb{C}, |w| < r, (1, w) \in \Lambda' \setminus \Lambda\}$.

First case: R is discrete. Since we have

$$\{(1, w) : |w| = 1\} \subset \Lambda',$$

there exists $0 < s < 1$ such that $s \leq |w| \leq 1$ implies $(1, w) \in \Lambda'$. Since R is discrete it must be at most countable, so we may choose s such that, further, $|w| = s$ implies

$(1, w) \in \Lambda$. Then $\{w: |w| < s, (1, w) \notin \Lambda\}$ is relatively compact in $B(0, s)$. There exists $t > 0$ such that the multifunction K, defined on $B(1, t)$ by

$$K(z) = \{w: w \in \mathbf{C}, |w| < s, (z, w) \notin \Lambda\},$$

is analytic. Suppose that $|z - 1| < t$, $|z| < 1$ and $w \in \partial K(z)$. Then $(z, w) \in \Lambda' \setminus \Lambda$, so it is an extension point for Λ, and consequently an extension point for the intersection of Λ and $B(1, t) \times B(0, s)$. By Theorem 7.2.1, w is a good isolated point for $K(z)$. But this holds for all $w \in \partial K(z)$, so $K(z)$ is finite for $|z - 1| < t$ and $|z| < 1$. By Theorem 7.1.7, $K(z)$ is finite for all z satisfying $|z - 1| < t$. In particular, $0 \in K(1)$ is a good isolated point so, by Theorem 7.2.1, $(1, 0) \in \Lambda'$. But this is a contradiction.

Second case: R is not discrete. We may choose a countable set of pairs (f_i, U_i), with the U_i open polydisks $N_i \times M_i$ and the f_i holomorphic and not identically zero on U_i, such that (z, w) is an extension point for Λ if and only if $f_i(z, w) = 0$ for some i. There exists δ with $0 < \delta < 1$ such that

$$|z| \leq 1 + \delta, |w| = 1 \text{ implies } (z, w) \in \Lambda'.$$

By Lemma 7.2.2, there exists $\eta \in \mathbf{C}$ with $|\eta| < \delta$ such that the sets

$$\{w: |w| < r, f_i(z + \eta w, w) = 0\} \tag{1}$$

are all discrete, for any $z \in \mathbf{C}$.

Let $J(z, w) = (z + \eta w, w)$ and $\Theta = J^{-1}(\Lambda)$. Because J is a biholomorphic mapping, Θ is pseudoconvex and $\Theta' = J^{-1}(\Lambda')$. Also since $|\eta| < \delta$, $|z| \leq 1$ and $|w| = 1$ imply $(z, w) \in \Theta'$. If we set

$$t_0 = \inf\{|z|: (z, w) \notin \Theta', \text{ for some } |w| \leq 1\},$$

then $0 < t_0 \leq 1$. Let z_0, w_0 be such that $|z_0| = t_0$, $|w_0| \leq 1$ and $(z_0, w_0) \notin \Theta'$. Then $|w_0| < 1$. So we have the following situation:

$$\Theta \text{ is pseudoconvex in } \mathbf{C} \times B(0, r),$$

$$|z| < t_0 \text{ and } |w| \leq 1 \text{ imply } (z, w) \in \Theta',$$

$$|z| \leq t_0 \text{ and } |w| = 1 \text{ imply } (z, w) \in \Theta',$$

$$(z_0, w_0) \notin \Theta' \text{ where } |z_0| = t_0, |w_0| < 1,$$

and by (1), $\{w : |w| < r, (z_0, w) \in \Theta' \setminus \Theta\}$ is discrete. Now using an argument similar to that of the first case, with $(0, 1)$ replaced by (z_0, w_0), we get a contradiction. $\square$

For more details on this proof, look at T.J. Ransford, *Analytic Multivalued Functions*, Doctoral Thesis, Cambridge, 1984.

By definition, we denote by $DK(\lambda)$ the set of points of $K(\lambda)$ which are not good isolated points. It is easy to prove that $DK(\lambda)$ is compact and satisfies $K(\lambda)' \subset DK(\lambda) \subset K(\lambda)$, where $K(\lambda)'$ denotes the set of limit points of $K(\lambda)$. By transfinite induction we can define $D^\alpha K(\lambda)$ for all ordinal numbers α by

$$D^\alpha K(\lambda) = D(D^{\alpha-1} K(\lambda)), \text{ if } \alpha \text{ is not a limit ordinal,}$$

$$D^\alpha K(\alpha) = \cap_{\beta < \alpha} D^\beta K(\lambda), \text{ if } \alpha \text{ is a limit ordinal,}$$

with the convention that $D^0 K(\lambda) = K(\lambda)$.

G. Cantor introduced the notion of an α-*derived set* of a closed set C defined by

$$C^{(\alpha)} = (C^{(\alpha-1)})', \text{ if } \alpha \text{ is not a limit ordinal,}$$

$$C^{(\alpha)} = \cap_{\beta < \alpha} C^{(\beta)}, \text{ if } \alpha \text{ is a limit ordinal,}$$

with the convention that $C^{(0)} = C$.

By transfinite induction it is easy to prove that $K(\lambda)^{(\alpha)} \subset D^\alpha K(\lambda)$, for all ordinal numbers α.

THEOREM 7.2.4 (K. OKA-T. NISHINO). *Let K be an analytic multifunction from a domain $D \subset \mathbf{C}$ into $\mathbf{C}$ and let α be an ordinal number. Then either $D^\alpha K(\lambda) \neq \emptyset$ for all $\lambda \in D$ and $D^\alpha K: \lambda \mapsto D^\alpha K(\lambda)$ is an analytic multifunction on D, or $D^\alpha K(\lambda) = \emptyset$ for all $\lambda \in D$. In the latter case let γ be the smallest ordinal such that $D^\gamma K(\lambda) = \emptyset$, for all $\lambda \in D$. Then γ is not a limit ordinal and there exist an integer n and a closed discrete subset F of D such that $\#D^{\gamma-1} K(\lambda) = n$, for $\lambda \in D \setminus F$, and $\#D^{\gamma-1} K(\lambda) \leq n - 1$, for $\lambda \in F$.*

PROOF. First we suppose that $\alpha = 1$. Let $\Omega = \{(\lambda, z): \lambda \in D, z \notin K(\lambda)\}$, which is pseudoconvex by Theorem 7.1.10. If Ω' denotes the union of Ω with its extension points then, by Theorem 7.2.3, Ω' is pseudoconvex. Let $L(\lambda) = \{\lambda: \lambda \in D, (\lambda, z) \notin \Omega'\}$. Then $L(\lambda) \subset K(\lambda)$, so L is locally bounded on D. By Theorem 7.1.10 applied to Ω', it is not difficult to see that either $L(\lambda) = \emptyset$ for all $\lambda \in D$ or $L(\lambda) \neq \emptyset$ for all $\lambda \in D$, in which case L is an analytic multifunction on D. It is easy to see that $\partial D K(\lambda) \subset \partial K(\lambda)$ so, by Theorem 7.2.1, we have $\partial D K(\lambda) \subset L(\lambda) \subset D K(\lambda)$, for all $\lambda \in D$. If $L(\lambda) \neq \emptyset$ for all $\lambda \in D$, then $D K(\lambda) \neq \emptyset$, for all $\lambda \in D$, and it is easy to see that DK is upper semicontinuous. Hence by Theorem 7.1.2, DK is an analytic multifunction on D. If $L(\lambda) = \emptyset$ for all $\lambda \in D$, then $\partial D K(\lambda) = \emptyset$ and consequently $DK(\lambda) = \emptyset$, for all $\lambda \in D$. This says that all points of $K(\lambda)$ are good isolated points of $K(\lambda)$. Hence $K(\lambda)$ is finite for all $\lambda \in D$, and we apply Theorem 7.1.7 to obtain the conclusion.

Suppose now that α is an arbitrary ordinal number. In order to use transfinite induction, we suppose that the previous properties have been proved for $D^\beta K$ with $\beta < \alpha$. If there exists $\beta < \alpha$ such that $D^\beta K(\lambda) = \emptyset$ for all $\lambda \in D$, it is obvious that $D^\alpha K(\lambda) = \emptyset$ for all $\lambda \in D$. So suppose that $D^\beta K(\lambda) \neq \emptyset$, for all $\beta < \alpha$ and all $\lambda \in D$. By hypothesis all the $D^\beta K$ are analytic multifunctions on D. If α is not a limit ordinal, by the first part of the proof $D^\alpha K(\lambda) = D(D^{\alpha-1} K(\lambda))$ is either non-empty and $D^\alpha K$ is an analytic multifunction on D, or $D^\alpha K(\lambda) = \emptyset$ for all $\lambda \in D$. In this latter case, $D^{\alpha-1} K(\lambda)$ is finite for all $\lambda \in D$ and we conclude as previously. If α is a limit ordinal then $\{\beta : \beta < \alpha\}$ is a countable well-ordered set such that $\beta < \beta' < \alpha$ implies $D^{\beta'} K(\lambda) \subset D^\beta K(\lambda)$. Thus $\cap_{\beta<\alpha} D^\beta K(\lambda)$ is non-empty and, by Theorem 7.1.1(i), $D^\alpha K$ is analytic multivalued on D. Using transfinite induction, we then get the result. $\square$

In fact we shall see in the next theorem that it is not necessary to consider $D^\alpha K(\lambda)$ for all $\alpha \geq \omega_1$, where ω_1 denotes the first uncountable ordinal number, because $D^\alpha K(\lambda)$ stabilizes after some $\gamma < \omega_1$. This was proved by B. Aupetit and J. Zemánek for $K(\lambda)$ at most countable, but the proof is the same in the general case.

THEOREM 7.2.5. *Let K be an analytic multifunction from $D \subset \mathbb{C}$ into $\mathbb{C}$. Then there exists $\gamma < \omega_1$ such that $D^\gamma K(\lambda) = D^\alpha K(\lambda)$ for all α such that $\gamma \leq \alpha < \omega_1$ and for all $\lambda \in D$.*

PROOF. By definition of $D^\alpha K(\lambda)$ and transfinite induction, it suffices to prove that there exists $\gamma < \omega_1$ such that $D^\gamma K(\lambda) = D^{\gamma+1} K(\lambda)$ for all $\lambda \in D$. First we fix $\lambda \in D$. By Kuratowski's theorem (see Theorem 5.7.9), there exists a smallest ordinal $\gamma(\lambda) < \omega_1$ such that $D^\alpha K(\lambda) = D^{\gamma(\lambda)} K(\lambda)$ for $\gamma(\lambda) \leq \alpha < \omega_1$. Let $E_\alpha = \{\lambda : \lambda \in D, \gamma(\lambda) > \alpha\}$. Then $E_\alpha = \cup_{\alpha \leq \beta < \omega_1} \{\lambda : \lambda \in D, D^\beta K(\lambda) \neq D^{\beta+1} K(\lambda)\}$. Now if $D^\beta K(\lambda_0) \neq D^{\beta+1} K(\lambda_0)$, then $D^\beta K(\lambda_0)$ contains a good isolated point z_0. This implies that $D^\beta K(\lambda)$ contains good isolated points for λ in a neighbourhood of λ_0. Consequently E_α is open. Then $F_\alpha = D \setminus E_\alpha$ is closed, and $\alpha < \beta$ implies $F_\alpha \subset F_\beta$. So, by Kuratowski's theorem, there exists $\gamma < \omega_1$ such that $\gamma \leq \alpha < \omega_1$ implies $E_\alpha = E_\gamma$. But $\lambda \notin E_{\gamma(\lambda)}$ for all $\lambda \in D$. If $\lambda \in E_\gamma$ then $\gamma(\lambda) > \gamma$, so $E_\gamma = E_{\gamma(\lambda)}$, and consequently $\lambda \in E_{\gamma(\lambda)}$, which is a contradiction. Thus $E_\gamma = \emptyset$, which means that $\gamma(\lambda) \leq \gamma$ for all $\lambda \in D$, and so $D^{\gamma+1} K(\lambda) = D^\gamma K(\lambda)$. $\square$

The classical Cantor-Bendixson theorem says that every closed subset of $\mathbf{C}$ is the disjoint union of a perfect set and an at most countable set. It can be generalized in the following form:

COROLLARY 7.2.6. *Let K be an analytic multifunction from a domain $D \subset \mathbf{C}$ into $\mathbf{C}$. Then for each $\lambda \in D$, $K(\lambda)$ is the disjoint union of two sets $L(\lambda)$, $M(\lambda)$ such that:*

(i) *either $L(\lambda) = \emptyset$, for all $\lambda \in D$ or L is an analytic multifunction from D into $\mathbf{C}$ such that $DL(\lambda) = L(\lambda)$, for all $\lambda \in D$,*

(ii) *$M(\lambda)$ is at most countable for all $\lambda \in D$.*

PROOF. Let $\gamma < \omega_1$ as in the statement of Theorem 7.2.5 and let $L(\lambda) = D^\gamma K(\lambda)$, $M(\lambda) = K(\lambda) \setminus L(\lambda)$. By Theorem 7.2.5 and the definition of γ, part (i) is true. For every $\lambda \in D$, $D^\alpha K(\lambda) \setminus D^{\alpha+1} K(\lambda)$ consists of isolated points, so it is at most countable. Then $L(\lambda)$ is a countable union of at most countable sets, hence it is at most countable. $\square$

THEOREM 7.2.7 (B.AUPETIT-J.ZEMÁNEK). *Let K be an analytic multifunction from $D \subset \mathbf{C}$ into $\mathbf{C}$ and let F be a closed subset of D having non-zero capacity. Suppose that $\lambda \in F$ implies $K(\lambda)$ at most countable. Then there exists $\lambda_0 \in F$ such that $DK(\lambda_0) \neq K(\lambda_0)$.*

PROOF. Let $(U_n)_{n \geq 1}$ be a countable base of $\mathbf{C}$. We introduce the set $F^c = F \setminus \cup \{F \cap U_n : c(F \cap U_n) = 0\}$. By Appendix A.1.21, $c(F^c) > 0$. Suppose that $DK(\lambda) = K(\lambda)$ for $\lambda \in F^c$. Then, by Theorem 7.1.7, $K(\lambda)$ is finite for all $\lambda \in D$, and so $DK(\lambda) = \emptyset$ which gives a contradiction. Hence there exists $\lambda_1 \in F^c$ such that $K(\lambda_1)$ is infinite. But $K(\lambda_1)$ is countable and so is not perfect, and consequently it contains an isolated point z_0. The same argument with $K(\lambda_1) \setminus \{z_0\}$, which is compact and countably infinite, shows that $K(\lambda_1)$ contains at least two isolated points z_0, z_1. For $i = 0, 1$ we choose open disks Δ_i with centres at z_i having disjoint closures and such that $K(\lambda_1) \cap \overline{\Delta}_i = \{z_i\}$. We then choose $0 < r < 1/2$ such that $\overline{B}(\lambda_1, r) \subset D$ and such that $|\lambda - \lambda_1| < r$ implies $K(\lambda) \cap \partial \Delta_i = \emptyset$, for $i = 0, 1$. Because $DK(\lambda_1) = K(\lambda_1)$, each z_i is isolated in $K(\lambda_1)$ but is not a good isolated point. Applying Theorems 7.1.5 and 7.1.7, we conclude that the two sets $E_i = \{\lambda : |\lambda - \lambda_1| < r, K(\lambda) \cap \Delta_i \text{ finite}\}$ have capacity zero. But $F^c \cap B(\lambda_1, r)$ does not have capacity zero, by definition of F, and $E_0 \cup E_1$ has capacity zero, so there exists $\lambda_2 \in F^c \cap B(\lambda_1, r)$ such that $K(\lambda_2) \cap \Delta_i$ is infinite, for $i = 0, 1$. As before we can find four distinct isolated points in $K(\lambda_1)$, say z_{00}, z_{01} in Δ_0 and z_{10}, z_{11} in Δ_1.

So take four open disks Δ_{ij} centred respectively at z_{ij}, having disjoint closures, such that

$$\Delta_{00} \cup \Delta_{01} \subset \Delta_0,$$

$$\Delta_{10} \cup \Delta_{11} \subset \Delta_1,$$

$$K(\lambda_2) \cap \overline{\Delta}_{ij} = \{z_{ij}\}.$$

By induction we can construct a sequence $\lambda_n \in F^c$ such that

(i) $|\lambda_{n+1} - \lambda_n| \le 2^{-n}$, $n = 1, 2, \ldots$,

(ii) $K(\lambda_n)$ contains at least 2^n distinct isolated points $z_{i_1 \ldots i_n}$ where i_k takes the values $0, 1$,

(iii) each $z_{i_1, \ldots, i_n}$ is the centre of an open disk $\Delta_{i_1, \ldots, i_n}$, with the property that all these 2^n disks have disjoint closures and $\Delta_{i_1, \ldots, i_{n-1}, i_n} \subset \Delta_{i_1, \ldots, i_{n-1}}$.

Now (λ_n) is a Cauchy sequence in the closed set F, so converges to some $\lambda_0 \in F$. To obtain a contradiction we show that $K(\lambda_0)$ is uncountable. Let $I = (i_1, i_2, \ldots)$ be an arbitrary sequence of 0s and 1s. Since $(z_{i_1}, z_{i_1 i_2}, z_{i_1 i_2 i_3}, \ldots)$ is bounded, it contains a subsequence converging to z_I which is in $K(\lambda_0)$ because K is upper semicontinuous. If $I \ne J$ then there exists a smallest integer k such that $i_k \ne j_k$, so we have

$$z_I \in \overline{\Delta}_{i_1, \ldots, i_{k-1} i_k}$$

$$z_J \in \overline{\Delta}_{i_1, \ldots, i_{k-1} j_k}$$

and the two disks are disjoint by construction, so $z_I \ne z_J$. But the set of sequences I is uncountable, so $K(\lambda_0)$ is infinite and uncountable. This is a contradiction. $\square$

THEOREM 7.2.8 (SCARCITY THEOREM FOR COUNTABLE ANALYTIC MULTIFUNCTIONS). *Let K be an analytic multifunction from a domain $D \subset \mathbf{C}$ into $\mathbf{C}$. Then either the set of λ, for which $K(\lambda)$ is at most countable, has capacity zero, or $K(\lambda)$ is at most countable for all $\lambda \in D$. In the latter situation there exists $\gamma < \omega_1$ such that $D^\gamma K(\lambda) = \emptyset$, for all $\lambda \in D$.*

PROOF. Suppose that the set of λ for which $K(\lambda)$ is at most countable does not have capacity zero. Then there exists a compact set $F \subset D$ such that $c(F) > 0$, and such that $K(\lambda)$ is at most countable for $\lambda \in F$. By Corollary 7.2.6, $K(\lambda)$ is the disjoint union of $L(\lambda)$ and $M(\lambda)$, with either $L(\lambda) = \emptyset$ for $\lambda \in D$, or L analytic multivalued with $DL(\lambda) = L(\lambda)$ for all $\lambda \in D$. In the latter situation $L(\lambda)$ is at most countable on F so, by Theorem 7.2.7, there exists $\lambda_0 \in F$ such that $DL(\lambda_0) \ne L(\lambda_0)$, which is a contradiction. So $L(\lambda) = D^\gamma K(\lambda) = \emptyset$, for all $\lambda \in D$. Moreover $K(\lambda) = M(\lambda)$ is at most countable on D. $\square$

REMARK.   This result is the best possible. Let F be a compact set having capacity zero. By Evans's theorem (Theorem A.1.24), there exists u subharmonic on $\mathbf{C}$, such that $F = \{\lambda: \lambda \in \mathbf{C}, u(\lambda) = -\infty\}$. We define the multifunction K by

$$K(\lambda) = \{z: z \in \mathbf{C}, |z| \leq e^{u(\lambda)}\}.$$

It is an analytic multifunction defined on $\mathbf{C}$ which satisfies $K(\lambda) = \{0\}$ on F and which is uncountable on $\mathbf{C} \setminus F$.

We now give the solution to the *General Pełczyński Conjecture* (first mentioned in [1], p. 86), the problem which was, in fact, the main motivation behind the introduction of analytic multifunctions.

THEOREM 7.2.9.   *Let A be a Banach algebra containing an absorbing set U such that $\operatorname{Sp} x$ is at most countable for all $x \in U$. Then the spectrum of every element of A is at most countable. If moreover A is separable, then it satisfies the properties of Theorem 5.7.9.*

PROOF.   The argument is similar to that at the beginning of the proof of Theorem 5.4.2, except that we use Theorem 7.2.8 instead of Theorem 3.4.25. ◻

COROLLARY 7.2.10.   *Let A be a Banach algebra with involution. Suppose that the real vector subspace H of self-adjoint elements contains an absorbing subset U such that $\operatorname{Sp} h$ is at most countable for all $h \in H$. Then every element of A has an at most countable spectrum.*

PROOF.   The argument is a slight modification of that used in the proof of Corollary 5.4.3, replacing Theorem 3.4.25 by Theorem 7.2.8. ◻

We now give an application of the Oka-Nishino theorem to the Identity Principle.

It is easy to see that Theorem 3.4.26 can be paraphrased, with an almost identical proof, in the following manner.

THEOREM 7.2.11.   *Let K be an analytic multifunction from a domain D of $\mathbf{C}$ into $\mathbf{C}$. Suppose that for all $\lambda \in D$ the set $K(\lambda)$ has at most 0 as a limit point. Let $z \neq 0$ be given. Then either $Z = \{\lambda: \lambda \in D, z \in K(\lambda)\}$ is a closed discrete subset of D or it is all D.*

The same argument even proves the following.

COROLLARY 7.2.12. *Let K be an analytic multifunction from D into $\mathbf{C}$ and let $z \in \mathbf{C}$ be fixed. Then every point of the set*

$$Z = \{\lambda \colon \lambda \in D, z \in K(\lambda) \setminus DK(\lambda)\}$$

is either isolated or interior.

If K is a countable analytic multifunction, the analogue of Theorem 7.2.1 cannot be true. For instance, let $K_0 = \{1/n \colon n = 1, 2, \ldots\} \cup \{0\}$ and let $K(\lambda) = \lambda + K_0$, which is an analytic multifunction on $\mathbf{C}$. Then $Z = \{\lambda \colon 1 \in K(\lambda)\}$ is neither discrete nor $\mathbf{C}$. Nevertheless we have the following.

THEOREM 7.2.13 (B.AUPETIT-J.ZEMÁNEK). *Let K be an at most countable analytic multifunction from a domain D of $\mathbf{C}$ into $\mathbf{C}$ and let $z \in \mathbf{C}$ be fixed. Then the set*

$$Z = \{\lambda \colon \lambda \in D, z \in K(\lambda)\}$$

is either at most countable or it is all D.

PROOF. By Theorem 7.2.8, there exists a smallest ordinal $\gamma < \omega_1$ such that $D^\gamma K(\lambda) = \emptyset$ for all $\lambda \in D$. Then $Z = \cup_{0 \le \alpha < \gamma} Z_\alpha$, where $Z_\alpha = \{\lambda \colon \lambda \in D, z \in D^\alpha K(\lambda) \setminus D^{\alpha+1} K(\lambda)\}$. By Corollary 7.2.12, applied to the analytic multifunction $D^\alpha K$, we conclude that $D^\alpha K$ has only isolated or interior points. Because the set of ordinals less than γ is countable, Z is the disjoint union of an open set and of an at most countable set. If the interior of Z is empty then Z is at most countable and we have finished. If not, we shall show that $Z = D$. First we note that Z is closed in D by upper semicontinuity of K, and so the boundary of Z in D is at most countable. Let F be the closure of the interior of Z in D. It is enough to prove that $F = D$. To do this we have to prove that F is open. Let $a \in F$ and let $r > 0$ be such that $\overline{B}(a, r) \subset D$. There exists b in the interior of E such that $|a - b| < r$. The set of half-lines Γ, with origin at b and such that $\Gamma \cap B(a, r)$ contains a boundary point of Z, is at most countable. So the interior of Z is dense in $B(a, r)$ and hence $F \supset B(a, r)$. $\square$

Obviously Theorem 7.2.13 is not true if K is a general analytic multifunction. For example if we take

$$K(\lambda) = \{z \colon |z| \le 1\} \cup \{z \colon |z| \le |\lambda|\},$$

then we have $2 \in K(\lambda)$ if and only if $|\lambda| \ge 2$.

An interesting problem would be to extend Theorem 7.2.13 to a larger class. For instance, suppose that K is an analytic multifunction from a domain D of $\mathbf{C}$

into $\mathbf{C}$ such that $K(\lambda)$ has capacity zero, for all $\lambda \in D$ (such an example is given by $\mathrm{Sp}\, f(\lambda)$, where the operators $f(\lambda)$ are *quasi-algebraic* in the sense of P. Halmos: see [2], pp. 251–253). Is it true that Z is a closed set of capacity zero? This problem is related to A. Sadullaev's result implying that the graph of K is a polar subset of $\mathbf{C}^2$. But is it completely polar? In a recent preprint entitled *Pseudoconcave sets and algebraic lemniscates* (in Russian), A. Sadullaev solves this last problem positively. If his arguments are correct then the first question also has a positive answer, but the proof is very difficult.

§3. Distribution of Values of Analytic Multifunctions

The famous theorem of Picard asserts that a non-constant entire function takes all the values of the complex plane except perhaps one point. But what happens for the union of all the spectral values of $f(\lambda)$ if f is an analytic function from $\mathbf{C}$ into $M_n(\mathbf{C})$? This problem was partly studied by E. Borel, G. Valiron and G. Rémoundos, but their arguments are not always very convincing (even H. Cartan gave some insights on the general situation, but with a false conclusion on the number of exceptional points). In the first part of this section we shall describe the work of A. Zraïbi on the solution of the previous problem with the help of Nevanlinna theory.

In the second part we intend to show the intimate connection between such analytic multifunctions and pseudoconvex open subsets of $\mathbf{C}^2$. This connection reduces many problems on analytic multifunctions — and hence many spectral problems — to purely geometrical problems on pseudoconvex sets. This geometrical idea gives a very simple proof of the generalization of Picard's theorem to arbitrary analytic multifunctions.

Let F be meromorphic for $|z| < R \leq +\infty$, and let $0 < r < R$. We define

$$m(r, F) = \frac{1}{2\pi} \int_0^{2\pi} \log^+ |F(re^{i\theta})|\, d\theta,$$

$$N(r, F) = \int_0^r \frac{n(t) - n(0)}{t}\, dt + n(0) \log r,$$

where $n(t)$ denotes the number of poles, with their multiplicity, in the disk $B(0, t)$, and

$$T(r, F) = m(r, F) + N(r, F).$$

R. Nevanlinna proved the following inequality:

LEMMA 7.3.1.   *If F is meromorphic for $|z| < R \le +\infty$, and if $0 \ne c_0 = F(0) \ne \infty$, then for $\rho < r < R$ we have*

$$m(\rho, \frac{F'}{F}) < 4\log^+ T(r, F) + 3\log^+ \frac{1}{r-\rho} + 4\log^+ r + 2\log^+ \frac{1}{\rho} + 4\log^+\log^+ \frac{1}{|c_0|} + 16.$$

Using the relation $\frac{F^{(k)}}{F} = \frac{F'}{F} \cdot \frac{F''}{F'} \cdots \frac{F^{(k)}}{F^{(k-1)}}$ and the previous lemma, it is possible to prove that there exist constant numbers $a_k, a'_k, b_k, b'_k, b''_k, c_k$ depending only on k such that

$$m\left(\rho, \frac{F^{(k)}}{F}\right) < a_k + a'_k \log^+\log^+ \left|\frac{1}{F(0)}\right| + b_k \log^+ \frac{1}{\rho} + b'_k \log \frac{r}{\rho} +$$

$$+ b''_k \log \frac{r}{r-\rho} + c_k \log^+ T(r, F).$$

See King-Lai Hiong, *Extension d'un théorème de M.R. Nevanlinna*, Paris, 1957.

The following lemma is a weak form of a theorem due to E. Borel.

LEMMA 7.3.2.   *Let $\phi_1, \ldots, \phi_n$ be n linearly independent entire functions such that $\phi_1 + \phi_2 + \cdots + \phi_n = 1$. Then at least one of the ϕ_i has a zero.*

SKETCH OF PROOF.   For all $k = 1, 2, \ldots, n-1$, we have $\phi_1^{(k)} + \cdots + \phi_n^{(k)} = 0$. The n functions ϕ_i being linearly independent, the Wronskian D is not identically zero, where

$$D = \begin{vmatrix} \phi_1 & \phi_2 & \cdots & \phi_n \\ \phi_1' & \phi_2' & \cdots & \phi_n' \\ \vdots & \vdots & \ddots & \vdots \\ \phi_1^{(n-1)} & \phi_2^{(n-1)} & \cdots & \phi_n^{(n-1)} \end{vmatrix}.$$

We also introduce

$$\Delta = \begin{vmatrix} 1 & 1 & \cdots & 1 \\ \frac{\phi_1'}{\phi_1} & \frac{\phi_2'}{\phi_2} & \cdots & \frac{\phi_n'}{\phi_n} \\ \vdots & \vdots & \ddots & \vdots \\ \frac{\phi_1^{(n-1)}}{\phi_1} & \frac{\phi_2^{(n-1)}}{\phi_2} & \cdots & \frac{\phi_n^{(n-1)}}{\phi_n} \end{vmatrix}.$$

Let D_i be the minor determinant corresponding to the i-th element in the first line of D, and let Δ_i be the corresponding minor determinant in Δ.

We have $\phi_1 = \frac{D}{\phi_2 \cdots \phi_n} : \frac{D}{\phi_1 \cdots \phi_n} = \frac{\Delta}{\Delta_1}$. Consequently $m(r, \phi_1) = m(r, \Delta/\Delta_1) < N(r, \Delta) - N(r, 1/\Delta) + m(r, \Delta) + m(r, \Delta_1) + O(1)$.

If all the ϕ_i are not vanishing, since $\Delta = D/\phi_2 \cdots \phi_n$, we get that $N(r, \Delta) - N(r, 1/\Delta) = N(r, D) - N(r, 1/D)$, and so

$$T(r, \phi_1) = m(r, \phi_1) < N(r, D) - N(r, 1/D) + m(r, \Delta_1) + m(r, \Delta) + O(1).$$

If $T(r)$ denotes the greatest of the $T(r, \phi_i)$, for $i = 1, \ldots, n$, then applying the previous inequalities we get

$$T(r) < O[\log T(r) + \log r],$$

except perhaps on some segments having a finite total length. So $\liminf_{r \to \infty} \frac{T(r)}{\log r} < +\infty$, which gives a contradiction, because this condition implies that the ϕ_i are polynomials. $\square$

Now we intend to generalize Picard's theorem to finite analytic multifunctions by exploiting an idea of G. Rémoundos (*Extension aux fonctions algébroïdes du théorème de M. Picard et ses généralisations*, Paris, 1938).

First we introduce the notion of spectral multiplicity.

LEMMA 7.3.3. *Let K be an analytic multifunction defined on an open subset D of $\mathbb{C}$ such that $K(\lambda)$ is finite for all λ in D. Let $K(\lambda_0) = \{\alpha_1, \ldots, \alpha_p\}$ and $\epsilon > 0$ be such that $B(\alpha_i, \epsilon) \cap B(\alpha_j, \epsilon) = \emptyset$ for $i \neq j$. Then there exist $\alpha > 0$ and integers $n_1, \ldots, n_p$ such that $\#(K(\lambda) \cap B(\alpha_i, \epsilon)) = n_i$ for $0 < |\lambda - \lambda_0| < \alpha$ and $i = 1, \ldots, p$.*

PROOF. By Theorem 7.1.7, there exist an integer n and a closed discrete subset F of D such that $\#K(\lambda) = n$ on $D \setminus F$ and $\#F(\lambda) < n$ on F. If $\lambda_0 \notin F$ the lemma is obvious. So we suppose that $\lambda_0 \in F$ and let $\eta > 0$ be such that $B(\lambda_0, \eta) \cap F = \{\lambda_0\}$ and $B(\lambda_0, \eta) \subset D$. Then $K(\lambda_0) = \{\alpha_1, \ldots, \alpha_p\}$ with $p < n$, and let $\epsilon > 0$ be such that $B(\alpha_i, \epsilon) \cap B(\alpha_j, \epsilon) \neq \emptyset$ for $i \neq j$. By the Localization Principle (Theorem 7.1.5), the multifunctions $\lambda \mapsto K(\lambda) \cap B(\alpha_i, \epsilon)$ are analytic multifunctions on a disk $B(\lambda_0, \alpha)$ with $\alpha < \eta$. So there exist integers n_i and closed discrete subsets F_i of $B(\lambda_0, \alpha)$ such that $\#(K(\lambda) \cap B(\alpha_i, \epsilon)) = n_i$ on $B(\lambda_0, \alpha) \setminus F_i$, $\#(K(\lambda) \cap B(\alpha_i, \epsilon)) < n_i$ on F_i, and $n_1 + \cdots + n_p = n$. By upper semicontinuity we may suppose that $K(\lambda)$ is included in $\cup_{i=1}^p B(\alpha_i, \epsilon)$ for $\lambda \in B(\lambda_0, \alpha)$, so we have $\#K(\lambda) = n$ for $\lambda \in B(\lambda_0, \alpha) \setminus \{\lambda_0\}$. $\square$

The integer n_i is called the *spectral multiplicity* of α_i.

LEMMA 7.3.4. *Let $F(\lambda, u) = u^n + A_1(\lambda)u^{n-1} + \cdots + A_n(\lambda)$ be defined on $\mathbf{C}^2$, where the $A_i(\lambda)$ are non-constant entire functions. Then F has at most $2n - 1$ exceptional values in the sense of Picard — that is, for every $u \in \mathbf{C}$ there exists $\lambda \in \mathbf{C}$ such that $F(\lambda, u) = 0$, except perhaps for at most $2n - 1$ values of u.*

PROOF. Let u_0 be an exceptional value. Then $F(\lambda, u_0)$ is entire and has no zeros. Consequently either $F(\lambda, u_0)$ is constant or $F(\lambda, u_0) = e^{h(\lambda)}$, for some non-constant entire function h.

First we suppose that $F(\lambda, u)$ has n distinct exceptional values $u_1, \ldots, u_n$ such that $F(\lambda, u_i) = k_i$, for all $\lambda \in \mathbf{C}$. The identities

$$u_1^n + A_1(\lambda)u_1^{n-1} + \cdots + A_n(\lambda) = k_1$$

$$\vdots \qquad \vdots \qquad \ddots \qquad \vdots \qquad \vdots$$

$$u_n^n + A_1(\lambda)u_n^{n-1} + \cdots + A_n(\lambda) = k_n$$

define a Cramer system for the unknowns $A_i(\lambda)$ because the determinant is a Vandermonde determinant

$$\begin{vmatrix} u_1^{n-1} & u_1^{n-2} & \cdots & u_1 & 1 \\ u_2^{n-1} & u_2^{n-2} & \cdots & u_2 & 1 \\ \vdots & \vdots & \ddots & \vdots & \vdots \\ u_n^{n-1} & u_n^{n-2} & \cdots & u_n & 1 \end{vmatrix} \neq 0.$$

Consequently $A_i(\lambda)$ is constant for all $i = 1, \ldots, n$, and this is a contradiction. Hence there exist at most $n - 1$ exceptional values $u_1, \ldots, u_{n+1}$ for which $F(\lambda, u_i)$ is constant.

Now we suppose that $v_0, v_1, \ldots, v_n$ are $n + 1$ distinct exceptional values such that $F(\lambda, v_i) = e^{h_i(\lambda)}$, where h_i is a non-constant entire function. We have the following identities:

$$A_1(\lambda)v_0^{n-1} + \cdots + A_n(\lambda) - e^{h_0(\lambda)} \qquad v_0^n$$

$$\vdots \qquad \ddots \qquad \vdots \qquad \vdots \qquad \vdots$$

$$A_1(\lambda)v_n^{n-1} + \cdots + A_n(\lambda) = e^{h_n(\lambda)} - v_n^n.$$

Let

$$q = \begin{vmatrix} v_0^n & v_0^{n-1} & \cdots & v_0 & 1 \\ v_1^n & v_1^{n-1} & \cdots & v_1 & 1 \\ \vdots & \vdots & \ddots & \vdots & \vdots \\ v_n^n & v_n^{n-1} & \cdots & v_n & 1 \end{vmatrix} \neq 0.$$

We have $q = \sum_{i=0}^{n}(-1)^i v_i^n q_i$, where

$$q_i = \begin{vmatrix} v_0^{n-1} & v_0^{n-2} & \cdots & v_0 & 1 \\ \vdots & \vdots & \ddots & \vdots & \vdots \\ v_{i-1}^{n-1} & v_{i-1}^{n-2} & \cdots & v_{i-1} & 1 \\ v_{i+1}^{n-1} & v_{i+1}^{n-2} & \cdots & v_{i+1} & 1 \\ \vdots & \vdots & \ddots & \vdots & \vdots \\ v_n^{n-1} & v_n^{n-2} & \cdots & v_n & 1 \end{vmatrix} \neq 0$$

is also a Vandermonde determinant. So we have

$$q = \sum_{i=0}^{n}(-1)^i q_i \left[e^{h_i(\lambda)} - \left(\sum_{k=1}^{n} A_k(\lambda) v_i^{n-k} \right) \right]$$

$$= \sum_{i=0}^{n}(-1)^i q_i e^{h_i(\lambda)} - \sum_{i=1}^{n}(-1)^i q_i \left(\sum_{k=1}^{n} A_k(\lambda) v_i^{n-k} \right)$$

$$= \sum_{i=0}^{n}(-1)^i q_i e^{h_i(\lambda)},$$

because the last term in the previous equation is

$$\begin{vmatrix} e^{h_0(\lambda)} - v_0^n & v_0^{n-1} & \cdots & v_0 & 1 \\ e^{h_1(\lambda)} - v_1^n & v_1^{n-1} & \cdots & v_1 & 1 \\ \vdots & \vdots & \ddots & \vdots & \vdots \\ e^{h_n(\lambda)} - v_n^n & v_n^{n-1} & \cdots & v_n & 1 \end{vmatrix} = 0.$$

Because $q_i \neq 0$, the determinant q can be written $\sum_{i=1}^{n} e^{g_i(\lambda)}$, and the functions e^{g_i} are linearly independent. So by Lemma 7.3.2, we get a contradiction. Hence F has at most n exceptional values of this type, and so $2n - 1$ exceptional values. $\square$

THEOREM 7.3.5. *Let K be a non-constant analytic multifunction on $\mathbb{C}$. Suppose that $K(\lambda)$ is finite on a set E having a non-zero capacity. Then there exists a smallest integer n such that $\#K(\lambda) \leq n$ for all $\lambda \in \mathbb{C}$ and $\mathbb{C} \setminus \cup_{\lambda \in \mathbb{C}} K(\lambda)$ has at most $2n - 1$ points.*

PROOF. The first part comes from theorem 7.1.7. So outside F we have $K(\lambda) = \{\alpha_1(\lambda), \ldots, \alpha_n(\lambda)\}$, where the α_i are locally holomorphic. Let

$$F(\lambda, u) = \prod_{i=1}^{n}(\alpha_i(\lambda) - u) = u^n + A_1(\lambda)u^{n-1} + \cdots + A_n(\lambda) \quad \text{for } \lambda \notin F.$$

This function can be extended analytically to all $\mathbf{C}^2$ by Lemma 7.3.3, counting each $\alpha_i(\lambda)$ with its multiplicity if $\lambda \in F$. The $A_i(\lambda)$ are well-defined in all $\mathbf{C}$, and they are entire because they can be expressed as symmetric functions of the α_i (in fact we use Radó's Extension Theorem). Moreover they are not all constant since K is not constant. So u is not in $\cup_{\lambda \in \mathbf{C}} K(\lambda)$ if and only if u is exceptional for F. We then apply Lemma 7.3.4. $\blacksquare$

This result is the best possible because, given arbitrary $2n - 1$ distinct points, it is possible to construct a finite analytic multifunction on $\mathbf{C}$ avoiding these points.

Let $a_1, \ldots, a_{2n-1}$ be given distinct points, and consider the following analytic function from $\mathbf{C}$ into $M_n(\mathbf{C})$ defined by

$$f(\lambda) = \begin{pmatrix} C_1 e^\lambda + a_1 & -C_2 e^\lambda & C_3 e^\lambda & \cdots & (-1)^n C_{n-1} e^\lambda & (-1)^{n+1} C_n e^\lambda \\ 1 & a_2 & 0 & \cdots & 0 & 0 \\ 0 & 1 & a_3 & \cdots & 0 & 0 \\ \vdots & \vdots & \vdots & \ddots & \vdots & \vdots \\ 0 & 0 & 0 & \cdots & 1 & a_n \end{pmatrix}.$$

We have

$$\det(f(\lambda) - z) = (a_1 - z) \cdots (a_n - z) + e^\lambda \left[\sum_{i=1}^{n-1} C_i (a_{i+1} - z) \cdots (a_n - z) + C_n \right].$$

Let

$$P(z) = \sum_{i=1}^{n-1} C_i (a_{i+1} - z) \cdots (a_n - z) + C_n.$$

By induction it is possible to choose the constants $C_1, \ldots, C_n$ such that we have $P(z) = (a_{n+1} - z) \cdots (a_{2n-1} - z)$, and consequently

$$\det(f(\lambda) - z) = (a_1 - z) \cdots (a_n - z) + e^\lambda (a_{n+1} - z) \cdots (a_{2n-1} - z).$$

Then the analytic multifunction $\lambda \mapsto \mathrm{Sp}\, f(\lambda) = \{z \colon \det(f(\lambda) - z) = 0\}$ avoids exactly the $2n - 1$ points $a_1, \ldots, a_{2n-1}$.

We shall now be interested in improving Picard's Theorem when the analytic multifunction takes an infinite number of values.

Theorem 3.4.14 and Theorem 3.4.15 can be paraphrased to obtain the following two results.

THEOREM 7.3.6.   *Let K be an analytic multifunction on $\mathbf{C}$ and let $E = \overline{\cup_{\lambda \in \mathbf{C}} K(\lambda)}$. Then the boundary of E is included in the boundary of $K(\lambda)$, for all λ. In particular, if K is bounded then $K(\lambda)\hat{\ }$ is constant.*

THEOREM 7.3.7.   *Let K be an analytic multifunction on $\mathbf{C}$. Then either $K(\lambda)\hat{\ }$ is constant or $\bigcup_{\lambda \in \mathbf{C}} K(\lambda)\hat{\ }$ is dense in $\mathbf{C}$.*

A. Zraïbi and the author obtained the following generalization of Picard's theorem to analytic multifunctions: if K is an analytic multifunction on $\mathbf{C}$, then either $K(\lambda)\hat{\ }$ is constant or the complement of the union of the sets $K(\lambda)\hat{\ }$ is a G_δ-set having zero capacity. The original proof uses Frostman's theorem and is rather complicated. We now intend to give an easy and more geometric proof. As explained in §1, Z. Słodkowski noticed that if K is an analytic multifunction then the complement of its graph is a pseudoconvex open subset of $\mathbf{C}^2$. Moreover, in an important paper (*Analytic set-valued functions and spectra*, Math. Ann. **256** (1981), pp. 363–386), he discovered the striking fact that the theory of analytic multifunctions with values in $\mathbf{C}$, the theory of fibres for uniform algebras, and spectral theory are locally equivalent. More precisely, if K is an analytic multifunction on an open subset U of $\mathbf{C}$, with values in $\mathbf{C}$, then for every relatively compact subdomain V of U there exists an analytic function from V into $\mathcal{L}(\ell^2)$ such that $K(\lambda) = \mathrm{Sp}\, f(\lambda)$, for all $\lambda \in V$, and there exist a uniform algebra A on a compact set K and two elements $f, g \in A$ such that $V \subset \mathbf{C} \setminus f(K)$ and $K(\lambda) = g(f^{-1}(\lambda))$ for all $\lambda \in V$.

The following lemma will show that it is always possible to associate a lot of analytic multifunctions to a pseudoconvex open subset of $\mathbf{C}^2$. So, in fact, the four theories of pseudoconvex open subsets of $\mathbf{C}^2$, analytic multifunctions with values in $\mathbf{C}$, spectra of analytic families of operators and fibres associated to uniform algebras are equivalent in the sense that any result in one of these theories will give new information in the other theories. This is one of the reasons why we believe that a deeper knowledge of the geometry of pseudoconvex open subsets of $\mathbf{C}^2$ will have a great impact on spectral theory.

LEMMA 7.3.8.   *Let Ω be a non-empty pseudoconvex open subset of $\mathbf{C}^2$ and let $(\lambda_0, a) \in \Omega$. Suppose that D is the open set of $\lambda \in \mathbf{C}$ such that $(\lambda, a) \in \Omega$. Then the multifunction K defined on D by*

$$K(\lambda) = \left\{ \frac{1}{z-a} + a \colon (\lambda, z) \notin \Omega \right\} \cup \{a\}$$

is analytic.

PROOF.   It is obvious that $K(\lambda)$ is non-empty for $\lambda \in D$. The set $K(\lambda)$ is closed because if $u_n \in K(\lambda)$, $\lim u_n = u$, then either $u = a$ (so $u \in K(\lambda)$), or $u \neq a$ and $\lim 1/(u_n - a) = 1/(u-a)$. But $(\lambda, a + 1/(u_n - a)) \notin \Omega$ implies $(\lambda, a + 1/(u-a)) \notin \Omega$, so $u \in K(\lambda)$. Moreover $K(\lambda)$ is a compact subset of $\mathbf{C}$ because if $u_n \in K(\lambda)$ with

$\lim |u_n| = +\infty$, it follows from $u_n = 1/(z_n - a)$, with $(\lambda, z_n) \notin \Omega$, that $\lim z_n = a$, so $(\lambda, a) \notin \Omega$, which is a contradiction.

Let us now show that $\Omega_1 = \{(\lambda, z): \lambda \in D, z \in \mathbf{C}, z \notin K(\lambda)\}$ is pseudoconvex. Because $a \in K(\lambda)$ for all $\lambda \in D$, we have $(\lambda, a) \notin \Omega_1$ for $\lambda \in D$, so

$$\Omega_1 = \left\{(\lambda, z): \lambda \in D, z \in \mathbf{C} \setminus \{a\}, \left(\lambda, \frac{1}{z - a} + a\right) \in \Omega\right\}$$
$$= u^{-1}(\Omega) \cap [D \times (\mathbf{C} \setminus \{a\})]$$

where $u(z) = a + 1/(z - a)$ is analytic on $D \times (\mathbf{C} \setminus \{a\})$. Thus Ω_1 is pseudoconvex because $u^{-1}(\Omega)$ and $D \times (\mathbf{C} \setminus \{a\})$ are pseudoconvex. $\square$

THEOREM 7.3.9. *Let Ω be a pseudoconvex open subset of $\mathbf{C}^2$ and let U be a domain of $\mathbf{C}$ such that $U \times \{0\} \subset \Omega$. Then we have the following properties:*

(i) *either the set of $\lambda \in U$ such that $U \times \{0\} \subset \Omega$ is a G_δ-set of capacity zero, or $U \times \mathbf{C} \subset \Omega$,*

(ii) *either the set of $\lambda \in U$ such that $\{\lambda\} \times \mathbf{C} \subset \Omega$, except for a finite number of points, is a G_δ-set of capacity zero, or $(U \times \mathbf{C}) \cap \Omega$ is the complement of an analytic variety.*

PROOF. This is obvious by applying Lemma 7.3.8 with $a = 0$, and Theorem 7.1.3, if we remark that $\{\lambda\} \times \mathbf{C} \subset \Omega$ is equivalent to $K(\lambda) = \{0\}$ and that $\{\lambda\} \times \mathbf{C} \subset \Omega$, except for a finite number of points, is equivalent to $K(\lambda)$ being finite. $\square$

We are now able to give a generalization of Picard's theorem to analytic multifunctions.

THEOREM 7.3.10 (PICARD THEOREM FOR ANALYTIC MULTIFUNCTIONS). *Let K be an analytic multifunction on $\mathbf{C}$. If U is a component of $\mathbf{C} \setminus K(\lambda_0)$, for some $\lambda_0 \in \mathbf{C}$, then either U is a component of $\mathbf{C} \setminus K(\lambda)$, for all $\lambda \in \mathbf{C}$, or $U \setminus \cup_{\lambda \in \mathbf{C}} K(\lambda)$ is a G_δ-set of capacity zero. In particular if we consider the analytic multifunction $K^\wedge$, then either $K(\lambda)^\wedge$ is constant or $\mathbf{C} \setminus \cup_{\lambda \in \mathbf{C}} K(\lambda)^\wedge$ is a G_δ-set of capacity zero. Moreover, if $K^\wedge$ is not constant and is not algebroid, then the set F of z for which $\{\lambda: z \in K(\lambda)^\wedge\}$ is finite, is a G_δ-set of capacity zero.*

PROOF. We may suppose that $\lambda_0 = 0$. Then $\Omega = \{(\lambda, z): z \notin K(\lambda)\}$ is pseudoconvex, so the same is true for $\Omega' = \{(z, \lambda): (\lambda, z) \in \Omega\}$. Moreover $U \times \{0\} \subset \Omega'$. So we get the first part by applying Theorem 7.3.9 (i). Considering $K(\lambda)^\wedge$, the only component U is the unbounded one, so by Theorem 7.3.6 and Theorem 7.3.9 (i) we get the result. We now prove the last part. Let $U = \mathbb{C} \setminus \cap_{\mu \in \mathbb{C}} K(\mu)^{\hat{}}$. Because the intersection of $\mathbb{C} \setminus K(\lambda)^{\hat{}}$ and $\mathbb{C} \setminus K(\mu)^{\hat{}}$ is always non-empty, U is connected. Moreover $F \subset U$. So by Theorem 7.3.9 (ii), we conclude that either F is a G_δ-set having zero capacity or $F = U$. In the last situation, using the argument given in the proof of Theorem 7.3.4, we conclude that $K^{\hat{}}$ is algebroid. $\square$

Is this result the best one? Given a compact set C of capacity zero, is it possible to construct an analytic multifunction K on $\mathbb{C}$ such that $\mathbb{C} \setminus \cup_{\lambda \in \mathbb{C}} K(\lambda)^{\hat{}} = C$? Is it even possible to do this for $K(\lambda) = \text{Sp} f(\lambda)$, where f is an analytic function from $\mathbb{C}$ into the algebra of compact operators on some Banach space? A. Zraïbi obtained the following particular cases:

— If C is a subset of $\mathbb{C}$ not containing 0 and having at most 0 as a limit point, then there exists an analytic multifunction K such that $\mathbb{C} \setminus \cup_{\lambda \in \mathbb{C}} K(\lambda)^{\hat{}} = C$ (see Exercise VII.12).

— If C is a compact subset of $\mathbb{C}$ of capacity zero then there exists an analytic multifunction K such that $\mathbb{C} \setminus \cup_{\lambda \in \mathbb{C}} K(\lambda) = C$ (but the problem is that $K(\lambda)$ has holes and the sets $K(\lambda)^{\hat{}}$ cover all the plane!).

It is interesting to note that Theorem 7.3.9 gives a new proof of Tsuji's theorem concerning the distribution of values of entire functions of two complex variables (M. Tsuji, *Potential Theory in Modern Function Theory*, Second Edition, New York, 1975). The original proof is complicated and uses conformal mapping.

THEOREM 7.3.11 (M. TSUJI). *Let $G(\lambda, \mu)$ be an entire function on $\mathbb{C}^2$ which is not of the form $G(\lambda, \mu) = e^{H(\lambda, \mu)}$, with H entire on $\mathbb{C}^2$. Then there exists a G_δ-set E having zero capacity such that for $\mu \notin E$ there exists λ in $\mathbb{C}$ satisfying $G(\lambda, \mu) = 0$. Moreover if G is not algebroid — that is there are no entire functions $a_1, \ldots, a_n$ such that $G(\lambda, \xi) = a_n(\mu)\lambda^n + \cdots + a_1(\mu)\lambda + a_0(\mu)$ — then there exists a G_δ-set F having zero capacity such that for $\mu \notin F$ there exist an infinite number of λ satisfying $G(\lambda, \mu) = 0$.*

PROOF. Let $U = \{\mu : G(0, \mu) \neq 0\}$. If $U = \emptyset$ then $G(0, \mu) = 0$, for all μ, so $G(\lambda, \mu) = \lambda^k H(\lambda, \mu)$, for some integer $k \geq 1$ and some entire function H for which $H(0, \mu) \neq 0$. So we can reduce the general situation to the case where $G(0, \mu) \neq 0$. In this case, $\mathbb{C} \setminus U$ is closed and discrete, so U is a domain. Let $\Omega = \{(\lambda, \mu) : G(\lambda, \mu) \neq 0\}$. This is a pseudoconvex open subset of $\mathbb{C}^2$ because it is the complement of an analytic variety. Moreover $\{0\} \times U \subset \Omega$. So, by Theorem 7.3.9, either the set of μ such that $\mathbb{C} \times \{\mu\} \subset \Omega$ is a G_δ-set of capacity zero, or $\mathbb{C} \times \{\mu\} \subset \Omega$ for all μ. Suppose that we are in this last situation. Then $G(\lambda, \mu) \neq 0$ for $\lambda \in \mathbb{C}$ and $\mu \in U$. In other words, $G(\lambda, \mu) = 0$ implies $\mu \in \mathbb{C} \setminus U$. But for μ fixed the set $\{\lambda \colon G(\lambda, \mu) = 0\}$ is either discrete or $G(\lambda, \mu) \equiv 0$ as a function of λ. The first case implies that the zeros of G are isolated but, by Hartogs's result, this is impossible for an entire function of two variables. If now $\mu \notin U$ implies $G(\lambda, \mu) = 0$, we can suppose for instance that $0 \notin U$, and so $G(\lambda, 0) \equiv 0$. If we write $G(\lambda, \mu) = \sum_{n=0}^{\infty} a_n(\mu)\lambda^n$ we conclude that $a_n(0) = 0$, so μ divides $a_n(\mu)$ for all n. Hence there exists a greatest integer k such that $G(\lambda, \mu) = \mu^k K(\lambda, \mu)$ with K entire and $K(\lambda, 0) \neq 0$. Then $K(\lambda, \mu)$ has isolated zeros on the line $\mathbb{C} \times \{0\}$, wich is a contradiction. So the first part of the theorem is proved. The proof of the last part is very similar to the proof of Theorem 7.3.10. $\square$

Given K an analytic multifunction on $\mathbb{C}$ and $0 \leq \alpha \leq 1$, it is easy to verify that $\lambda \mapsto \alpha K(\lambda) + (1 - \alpha)K(\lambda)$ is analytic. This implies that $\lambda \mapsto \operatorname{co} K(\lambda) = \cup_{0 \leq \alpha \leq 1}[\alpha K(\lambda) + (1 - \alpha)K(\lambda)]$ is also an analytic multifunction on $\mathbb{C}$ (see Exercise VII.4). For convex analytic multifunctions it is possible to improve Theorem 7.3.10.

THEOREM 7.3.12. *Let K be a non-constant convex analytic multifunction on $\mathbb{C}$. Suppose that $\cap_{\lambda \in \mathbb{C}} K(\lambda) \neq \emptyset$. Then $\cup_{\lambda \in \mathbb{C}} K(\lambda) = \mathbb{C}$.*

PROOF. Let $a \in \cap_{\lambda \in \mathbb{C}} K(\lambda)$ and $z \in \mathbb{C}$. The half-line with origin at a containing z has non-zero capacity, so by Theorem 7.3.10 there exists $b \in K(\lambda_0)$ such that z is on the segment $[a, b]$. But $[a, b] \subset K(\lambda_0)$, so $z \in K(\lambda_0)$. $\square$

As a corollary we immediately obtain the following result of J.P. Williams.

COROLLARY 7.3.13. *Let a,b be two non-commuting elements of a Banach algebra A. We define $W(x) = \{f(x) \colon f \in A', \|f\| = f(1) = 1\}$ to be the numerical range of X. Then $\cup_{\lambda \in \mathbb{C}} W(e^{\lambda b} a e^{-\lambda b}) = \mathbb{C}$.*

PROOF. The multifunction $\lambda \mapsto W(e^{\lambda b} a e^{-\lambda b})$ is trivially analytic because it has entire selections. Morevover, it is not constant because $ab \neq ba$, and $\operatorname{Sp} a \subset \cap_{\lambda \in \mathbb{C}} W(e^{\lambda b} a e^{-\lambda b})$. So we get the result by Theorem 7.3.12. $\square$

In fact these last results are particular cases of the following result, which was obtained by T.J. Ransford with the help of covering spaces and lifts of multifunctions.

THEOREM 7.3.14. *Let K be an analytic multifunction on $\mathbb{C}$ and suppose that $K(\lambda)$ is connected for all $\lambda \in \mathbb{C}$. Then either $K(\lambda)\hat{\ }$ is constant or the union of all $K(\lambda)\hat{\ }$ covers all the plane except perhaps one point.*

For the proof, see the thesis of T.J. Ransford.

This result implies in particular that for a convex analytic multifunction defined on the complex plane, only one of the following possibilities occurs:

(i) if $\cup_{\lambda \in \mathbf{C}} K(\lambda)$ avoids two points of $\mathbf{C}$, then $K(\lambda)$ is constant on $\mathbf{C}$,

(ii) if $\cup_{\lambda \in \mathbf{C}} K(\lambda)$ avoids one point $\alpha \in \mathbf{C}$ then $K(\lambda)$ has the form $K(\lambda) = \alpha + e^{h(\lambda)} K_0$, where h is an entire function and K_0 is a fixed compact convex set,

(iii) $\cup_{\lambda \in \mathbf{C}} K(\lambda) = \mathbf{C}$.

§4. Conclusion

We hope that this quick introduction to analytic multifunctions has inspired the reader to learn more on this new field. Further material can be found in the publications of H. Alexander & J. Wermer, B. Aupetit, B. Aupetit & J. Zemánek, B. Aupetit & A. Zraïbi, T.J. Ransford, Z. Słodkowski, and J. Wermer. In these, many applications to spectral theory and to the theory of uniform algebras are given.

Other applications of great interest have been given to the theory of spectral interpolation (T.J. Ransford) and, very surprisingly, to the Corona Problem mentioned in Chapter IV, §1 (see B. Berndtsson and T.J. Ransford, *Analytic Multifunctions, the $\overline{\partial}$-equations and a proof of the Corona Problem*, Pacific J. Math. 1986; Z. Słodkowski, *An analytic set-valued selection and its applications to the corona theorem*, Trans. Amer. Math. Soc. 1986; and Z. Słodkowski, *On bounded analytic functions in finitely connected domains*, Trans. Amer. Math. Soc. 1987).

In order to generalize the quantum mechanical formalism, P. Jordan, J. von Neumann and E. Wigner introduced, in 1934, the notion of what is now called a *Jordan algebra*. This theory has been intensively studied since then (see for instance the recent book by K.A. Zhevlakov, A.M. Slin'ko, I.P. Shestakov, A.I. Shirshov, *Rings That Are Nearly Associative*, New York, 1982). Several mathematicians studied, with some success, the theory of complete normed Jordan algebras, that is the *Jordan-Banach algebras* (see for instance the important paper of E.M. Alfsen, F.W. Shultz and E. Størmer, *A Gelfand-Neumark theorem for Jordan algebras*, Adv. in Math. 1978). The absence of associativity in Jordan-Banach algebras implies great difficulties for calculations. But in the last five years the use of subharmonic methods and of the theory of analytic multifunctions has given a great number of important results in this non-associative theory. Obviously we shall not give any details, but the reader will find these new methods and results in the publications of B. Aupetit, B. Aupetit & L. Baribeau, B. Aupetit & M.A. Youngson, B. Aupetit & A. Zraïbi, M. Benslimane & A. Kaïdi & A. Rodríguez Palacios.

EXERCISE 1.   Let K be an upper semicontinuous multifunction from $D \subset \mathbf{C}$ into $\mathbf{C}$, having holomorphic selections. Prove that K is a continuous analytic multifunction on D.

EXERCISE 2.   Let f be an analytic function from $D \subset \mathbf{C}$ into $M_n(\mathbf{C})$. Prove that $\lambda \mapsto \operatorname{Sp} f(\lambda)$ is an analytic multifunction on D (Let D' be the set of $\lambda \in D$ such that $\det(z - f(\lambda))$ has distinct roots. Using the Implicit Function Theorem prove the result on D' and then, by continuity, extend it to D).

EXERCISE 3.   Let (K_n) be a sequence of analytic multifunctions defined on $D \subset \mathbf{C}^n$ with values in $\mathbf{C}^m$. Suppose that for each $\lambda \in D$ there exists a non-empty compact subset $K(\lambda)$ of $\mathbf{C}^m$ such that

$$\lim_{n \to \infty} \Delta(K_n(\lambda), K(\lambda)) = 0,$$

uniformly on each compact subset of D. Prove that K is an analytic multifunction on D, which is continuous if the K_n are continuous.

EXERCISE 4.   Let K_1, K_2 be two analytic multifunctions from $D \subset \mathbf{C}^n$ into $\mathbf{C}^m$. Prove that $K_1 + K_2$ is an analytic multifunction. Conclude that the convex hull of an analytic multifunction is also an analytic multifunction.

EXERCISE 5.   Let K be an analytic multifunction from $D \subset \mathbf{C}^n$ into $\mathbf{C}$ and let $\Omega = \{(\lambda, z) : \lambda \in D, z \notin K(\lambda)\}$ be the complement of its graph. Prove that $(\lambda, z) \mapsto -\log \operatorname{dist}(z, K(\lambda))$ is plurisubharmonic on Ω.

EXERCISE 6.   Extend Theorems 3.4.11–3.4.19 to the situation of analytic multifunctions.

EXERCISE 7.   Let K be an analytic multifunction from $D \subset \mathbf{C}$ into $\mathbf{C}$. Suppose that, for some $\lambda_0 \in D$, $K(\lambda_0)$ is totally disconnected. Prove that K is continuous at λ_0.

*EXERCISE 8.   With the notation of Theorem 7.1.14, prove that the function $\lambda \mapsto \log \rho(g(f^{-1}(\lambda)))$ is subharmonic on W.

**EXERCISE 9.   Try to prove Theorem 7.1.14.

*EXERCISE 10.   Using Theorem 7.1.14 and Theorem 7.1.7, give a new proof of E. Bishop's theorem on analytic structure ([10], Theorem 11.2). Prove that it is even possible to replace the hypothesis "G of positive planar measure" by the weaker hypothesis "G of positive capacity".

*EXERCISE 11. Using Theorem 7.1.14 and Theorem 7.2.8, give a new proof of R. Basener's theorem on analytic structure ([10], Theorem 20.3). Prove that it is even sufficient to suppose that the fibres are at most countable on a subset G of W having positive capacity.

*EXERCISE 12. Let C be a subset of the complex plane not containing 0 and having at most 0 as a limit point. Prove that there exists an analytic multifunction K defined on $\mathbf{C}$ with values in $\mathbf{C}$, such that $K(\lambda)$ has at most 0 as a limit point for each $\lambda \in \mathbf{C}$, and satisfying

$$\mathbf{C} \setminus U_{\lambda \in \mathbf{C}} K(\lambda) = C.$$

*EXERCISE 13. Let K be an analytic multifunction from a domain D of $\mathbf{C}$ into $\mathbf{C}$. Suppose that $K(\lambda)$ is convex and never contains 0 for $\lambda \in D$. Denote by $\alpha(\lambda)$ the angle under which $K(\lambda)$ is seen from 0. Prove that α is subharmonic on D. If moreover $D = \mathbf{C}$, conclude that $K(\lambda)$ has the form

$$K(\lambda) = e^{h(\lambda)} K_0$$

where h is an entire function and K_0 is a fixed compact and convex set not containing 0.

*EXERCISE 14. Let K be an analytic multifunction from an open set D of $\mathbf{C}$ into $\mathbf{C}$.

(i) Suppose first that $K(\lambda)$ is a segment $[0, \alpha(\lambda)]$, for each $\lambda \in D$. Prove that α is holomorphic on D.

(ii) Suppose now that K is continuous on D and that $K(\lambda)$ is a segment $[\alpha(\lambda), \beta(\lambda)]$, for each $\lambda \in D$. Prove that $\alpha + \beta$ and $\alpha\beta$ are holomorphic on D. Conclude that there exists an analytic function f from D into $M_2(\mathbf{C})$ such that $K(\lambda) = \mathrm{co}\,\mathrm{Sp}\,f(\lambda)$, for every $\lambda \in D$.

**EXERCISE 15. Let K be an analytic multifunction from an open set D of $\mathbf{C}$ into $\mathbf{C}$. Suppose that $K(\lambda)$ has at most zero as a limit point, for each $\lambda \in D$. Does there exist an analytic function f from D into the algebra of compact operators on ℓ^2, such that $K(\lambda) = \mathrm{Sp}\,f(\lambda)$, for each $\lambda \in D$?

APPENDIX

This appendix is essentially a list of results without proofs. The reader intending to learn more about the deep results given in these two sections will be obliged to refer to the selection of books mentioned below.

§1. Subharmonic Functions and Capacity

Unfortunately, the most important results on subharmonic functions and capacity are scattered in many books, for instance in [4,5,9] and in the following books: M. Brelot, *Eléments de la théorie classique du potentiel*, Paris, 1965; R. Narasimhan, *Several Complex Variables*, Chicago, 1971; I. Privaloff, *Subharmonic Functions* (in Russian), Moscow, 1937; T. Radó, *Subharmonic Functions*, New York, 1949; M. Tsuji, *Potential Theory in Modern Function Theory*, New York 1975. For a brief survey, look at the appendix of [1].

THEOREM A.1.1.   *Let D be an open subset of the complex plane.*

 (i) *If ϕ_1 and ϕ_2 are subharmonic on D then $\phi_1 + \phi_2$ is subharmonic on D.*

 (ii) *If ϕ is subharmonic on D and if α is a positive number then $\alpha \cdot \phi$ is subharmonic on D.*

 (iii) *If ϕ_1 and ϕ_2 are subharmonic on D then $\max(\phi_1, \phi_2)$ is subharmonic on D.*

 (iv) *If ϕ is subharmonic on D and if f is a real, convex and increasing function on $\mathbf{R}$, then $f \circ \phi$ is subharmonic on D (by convention $f(-\infty) = \lim f(x)$ when x goes to $-\infty$).*

 (v) *If (ϕ_n) is a sequence of subharmonic functions on D converging uniformly to ϕ on each compact subset of D, then ϕ is subharmonic on D.*

 (vi) *If (ϕ_n) is a decreasing sequence of subharmonic functions on D then $\phi = \lim \phi_n$ is subharmonic on D.*

(vii) *If ϕ is locally integrable on D, and satisfies the mean inequality, and if $\phi^*(z) = \limsup\{\phi(u): u \to z\} \geq \phi(z)$, called the upper regularization of ϕ, satisfies $\phi^*(z) < +\infty$ for all z in D, then ϕ^* is subharmonic on D.*

(viii) *Let K be a compact set and μ be a positive measure on K. Suppose that $(\phi_t)_{t \in K}$ is a family of subharmonic functions on D such that $(t, z) \mapsto \phi_t(z)$ is measurable on $K \times \partial B(a, r)$ for all $\overline{B}(a, r) \subset D$, and such that $t \mapsto \phi_t(z)$ is μ-integrable for all $z \in D$. Then $\phi(z) = \int_K \phi_t(z)\,d\mu(t)$ is subharmonic on D.*

For ϕ subharmonic on D, $a \in D$ and $r > 0$ such that $\overline{B}(a, r) \subset D$, we introduce the following notation:

$$N(a, r, \phi) = \frac{1}{2\pi}\int_0^{2\pi} \phi(a + re^{i\theta})d\theta\ , \quad M(a, r, \phi) = \max_{0 \leq \theta \leq 2\pi} \phi(a + re^{i\theta}).$$

THEOREM A.1.2. *If ϕ is subharmonic on an open set D, then for every a in D we have*

$$\phi(a) = \limsup_{\substack{z \to a \\ z \neq a}} \phi(z) = \lim_{\substack{r \to 0 \\ r > 0}} N(a, r, \phi) = \lim_{\substack{r \to 0 \\ r > 0}} \overset{\circ}{M}(a, r, \phi).$$

THEOREM A.1.3 (MAXIMUM PRINCIPLE FOR SUBHARMONIC FUNCTIONS). *Let ϕ be a subharmonic function on a domain D. Suppose that there exists a in D such that $\phi(z) \leq \phi(a)$ for all z in D. Then $\phi(z) = \phi(a)$ for all z in D.*

COROLLARY A.1.4. *Let D be a bounded domain and ϕ be subharmonic on D. Suppose that there exists M such that for every $\xi \in \partial D$ we have $\limsup_{\substack{z \to \xi \\ z \in D}} \phi(z) \leq M$. Then we have $\phi(z) < M$ on D or ϕ is constant on D.*

If D is unbounded, the same result is true if we suppose further that $\limsup_{\substack{z \to \infty \\ z \in D}} \phi(z) \leq M$.

THEOREM A.1.5. *Let ϕ be upper semicontinuous on an open subset D of $\mathbf{C}$. Then ϕ is subharmonic on D if and only if, for every closed disk included in D and for every polynomial p, the inequality $\phi(z) \leq \operatorname{Re} p(z)$ on the boundary of the disk implies the same inequality on all the disk.*

COROLLARY A.1.6. *Let ϕ_1, ϕ_2 be two positive functions such that $\log \phi_1$ and $\log \phi_2$ are subharmonic on an open set D. Then $\log(\phi_1 + \phi_2)$ is subharmonic on D.*

COROLLARY A.1.7. *Let D be an open subset of* C. *Then* $-\log \text{dist}(\lambda, \partial D)$ *is subharmonic on D.*

THEOREM A.1.8 (E.F.BECKENBACH-S.SAKS). *Let ϕ be positive on an open set D. Then $\log \phi$ is subharmonic on D if and only if $z \mapsto |e^{p(z)}|\phi(z)$ is subharmonic on D for every polynomial p.*

THEOREM A.1.9 (LOCAL CHARACTERIZATION OF SUBHARMONIC FUNCTIONS).
Suppose that ϕ is upper semicontinuous on an open set D and suppose that for each a in D there exists a sequence (r_n) satisfying $r_n > 0$, $\lim_{n \to \infty} r_n = 0$ and

$$\phi(a) \leq \frac{1}{2\pi} \int_0^{2\pi} \phi(a + r_n e^{i\theta}) d\theta,$$

for all $n = 1, 2, \ldots$ Then ϕ is subharmonic on D. In particular ϕ is subharmonic on D if and only if it is locally subharmonic.

COROLLARY A.1.10. *If ϕ is of class C^2 on an open set D then ϕ is subharmonic on D if and only if $\Delta\phi$ is positive on D.*

THEOREM A.1.11 (LIOUVILLE'S THEOREM FOR SUBHARMONIC FUNCTIONS). *If ϕ is subharmonic on the complex plane and if $\liminf_{r \to \infty} \frac{M(0,r,\phi)}{\log r} = 0$, then ϕ is constant.*

THEOREM A.1.12 (T. RADÓ). *Let ϕ be a subharmonic function on an open set D. There exists an increasing sequence of open sets D_n which exhaust D and are relatively compact in D, and a decreasing sequence of functions ϕ_n subharmonic on D_n such that $\phi_n \in C^\infty(D_n)$ and $\phi(z) = \lim_{n \to \infty} \phi_n(z)$ on D.*

THEOREM A.1.13. *Let h be holomorphic on an open set D and ϕ be subharmonic on a neighbourhood of $h(D)$. Then $\phi \circ h$ is subharmonic on D.*

Using Theorem A.1.12 it is possible to obtain the following improvement of Theorem A.1.8.

THEOREM A.1.14 (P.MONTEL-T.RADÓ). *Let ϕ be positive on an open set D. Then $\log \phi$ is subharmonic on D if and only if $z \mapsto |e^{\alpha z}|\phi(z)$ is subharmonic on D, for all complex numbers α.*

The following theorem, which is an important step in the proof of H. Cartan's theorem, uses Theorem 1.1.7 in its proof.

THEOREM A.1.15 (F. RIESZ).   *Suppose that ϕ is subharmonic and not identical to $-\infty$ on a domain D. Then there exists a unique Borel positive measure μ on D such that for every compact subset K of D and every z in D, we have*

$$\phi(z) = \int_K \log|z - \xi|\, d\mu(\xi) \ + \ h(z),$$

where h is harmonic on the interior of K.

We now study the notions of thin and non-thin sets, first introduced by M. Brelot.

We say that a subset E of $\mathbf{C}$ is *non-thin at a* if $a \in \overline{E}$ and if for all ϕ subharmonic on a neighbourhood of a we have

$$\phi(a) = \limsup_{\substack{z \to a \\ z \neq a \\ z \in E}} \phi(z).$$

For instance, by Theorem A.1.2, $\mathbf{C} \setminus \{a\}$ is non-thin at a. If $a \in \overline{E} \setminus E$, then E is *thin at a*, if there exists ϕ subharmonic in a neighbourhood of a such that $\phi(a) > \limsup_{\substack{z \to a \\ z \neq a \\ z \in E}} \phi(z)$.

It is easy to verify that E non-thin at a and $E \subset F$ implies F non-thin at a.

THEOREM A.1.16 (M. BRELOT).   *Let K be a compact subset of an open set D in $\mathbf{C}$. Suppose that there exists ϕ subharmonic on $D \setminus K$, such that $\phi(z) < 0$ on $D \setminus K$ and $\lim_{\substack{z \to a \\ z \in D \setminus K}} \phi(z) = 0$ for some $a \in \partial K$. Then K is non-thin at a.*

THEOREM A.1.17 (K. OKA-W. ROTHSTEIN).   *If Γ is an arc in $\mathbf{C}$ then it is non-thin at each of its points.*

THEOREM A.1.18.   *Let U be a domain in $\mathbf{C}$. Then U is non-thin at each of its boundary points.*

We now introduce the notion of *capacity* for Borel subsets of the complex plane, which is in some sense a measure of their size. First we begin with compact sets. For K compact and μ a positive measure supported by K, the *logarithmic potential* is defined by

$$u(z) = \int_K \log|z - \xi|\, d\mu(\xi).$$

Then u is subharmonic on the complex plane and harmonic outside of K. We suppose now that μ is a probability measure concentrated on K. For $z \in K$ we have $\log |z - \xi| \leq \log \delta(K)$, where $\delta(K)$ denotes the diameter of K. Consequently,

$$I(\mu) = \int_K u(z)\, d\mu = \iint_{K \times K} \log |z - \xi|\, d\mu(\xi) d\mu(z)$$

exists and is bounded above by $\log \delta(K)$. We now introduce the *equilibrium value* $V(K)$ of K, given by

$$-\infty \leq V(K) = \sup_\mu I(\mu) \leq \log \delta(K)$$

for all the probability measures μ concentrated on K. The *capacity* of K is $c(K) = e^{V(K)}$. Of course $0 \leq c(K) \leq \delta(K)$. This quantity is difficult to calculate but we shall give other geometrical definitions latter. It is obvious that $c(K_1) \leq c(K_2)$ if $K_1 \subset K_2$.

THEOREM A.1.19.   *Suppose that K is a compact subset of $\mathbb{C}$ such that $V(K) > -\infty$, that is $c(K) > 0$. Then there exists a unique probability measure μ supported by K such that $I(\mu) = V(K)$.*

We now extend the definition of capacity to any subset of $\mathbb{C}$. If $E \subset \mathbb{C}$ we define the *inner capacity* of E by

$$c^-(E) = \sup_{K \subset E} c(K) \leq +\infty$$

for all compact subsets K of E, and we define the *outer capacity* of E by

$$c^+(E) = \inf_{U \supset E} c^-(U) \leq +\infty$$

for all open sets U containing E.

The following properties are easy to prove:

(i) for every $E \subset \mathbb{C}$ we have $c^-(E) \leq c^+(E)$,

(ii) if $E \subset F$ then $c^-(E) \leq c^-(F)$ and $c^+(E) \leq c^+(F)$,

(iii) if E' is the image of E by the transformation $z \mapsto \alpha z + \beta$ then $c^-(E') = |\alpha| c^-(E)$ and $c^+(E') = |\alpha| c^+(E)$,

(iv) if U is open then $c^+(U) = c^-(U)$.

We say that a set E is *capacitable* if $c^-(E) = c^+(E)$. In that case we denote by $c(E)$ this common value. Of course open sets are capacitable. In the next theorem we obtain that compact sets are capacitable. A very deep result of G. Choquet tells us that bounded Borel sets, more generally bounded analytic sets, are capacitable.

THEOREM A.1.20. *Let K be a compact subset of $\mathbb{C}$ and (U_n) be a sequence of bounded open sets such that $\overline{U}_{n+1} \subset U_n$ and $K = \cap_{n=1}^{\infty} U_n$. Then $c^-(K) = \lim_{n \to \infty} c(U_n)$, and consequently K is capacitable.*

THEOREM A.1.21. *If (E_n) is a sequence of Borel sets with inner capacity zero then their union is a Borel set of inner capacity zero.*

We denote by $\mathfrak{P}_n$ the set of polynomials of the form $p(z) = z^n + a_1 z^{n-1} + \cdots + a_n$. Also, we denote by $t_n(K)$ the quantity $\inf_{p \in \mathfrak{P}_n} \max_{z \in K} |p(z)|$. A compactness argument in the finite-dimensional vector space $\mathfrak{P}_n$ shows that $t_n(K) = \max_{z \in K} |p_n(z)|$ for some $p_n \in \mathfrak{P}_n$, which is in fact unique for K given and is called the *n-th Tchebycheff polynomial* of K.

THEOREM A.1.22 (M.FEKETE-G.SZEGÖ). *For any compact subset K of $\mathbb{C}$ we have*

$$c(K) = \lim_{n \to \infty} \delta_n(K) = \lim_{n \to \infty} (t_n(K))^{1/n}.$$

COROLLARY A.1.23. *For any compact subset K of $\mathbb{C}$ we have $c(K) = c(\partial_e K)$, where $\partial_e K$ denotes the outer boundary of K.*

THEOREM A.1.24 (G.C. EVANS). *Let K be a compact subset of $\mathbb{C}$ having zero capacity. There exists a probability measure μ, supported by K, such that, for the corresponding logarithmic potential $u(z) = \int_K \log |z - \xi| \, d\mu(\xi)$, we have $K = \{z : u(z) = -\infty\}$.*

THEOREM A.1.25. *Let K be a compact and connected subset of $\mathbb{C}$. We denote by D the unbounded component of $\mathbb{C} \setminus K$. Suppose that there exists a conformal mapping w from D onto $\{z : |z| > R\}$ that is continuously extendable to the boundary and such that*

$$w(z) = z + a_0 + \frac{a_1}{z} + \frac{a_2}{z^2} + \cdots, \quad \text{near infinity.}$$

Then $c(K) = R$.

COROLLARY A.1.26. *If K is a closed disk of radius R then $c(K) = R$.*

COROLLARY A.1.27. *If I is a closed segment of length L then $c(I) = L/4$.*

We now see that compact sets having zero capacity are very small in some sense.

Let h be an increasing function on $[0, 1]$ such that $h(0) = 0$. Given a bounded subset E of $\mathbf{C}$ we consider all the coverings of E by a finite or countable number of squares having sides parallel to the coordinate axes and length of the sides less than ϵ. We put $H^\epsilon(E) = \inf \sum h(d_i) \le +\infty$, for all the coverings $E \subset \cup_{i=1}^\infty C_i$ having the previous properties, d_i denoting the length of the sides of C_i. If ϵ decreases to zero then $H^\epsilon(E)$ increases. By definition,

$$H(E) = \lim_{\epsilon \downarrow 0} H^\epsilon(E) \le +\infty.$$

This quantity is the *h-Hausdorff measure* of E. If $h(t) = t^\alpha$, with $\alpha > 0$, then the corresponding quantity $H_\alpha(E)$ is the *α-Hausdorff measure* of E. If $\alpha = 2$ it is exactly the outer Lebesque planar measure. If $\alpha = 1$ it is the outer linear measure. Obviously, because E is bounded, we have $H_2(E) < +\infty$. It is not difficult to prove that $H_\alpha(E) < +\infty$ implies $H_\beta(E) = 0$ for $\beta > \alpha$.

THEOREM A.1.28. *If K is a compact subset of $\mathbf{C}$ having zero capacity then it is totally disconnected and $H_\alpha(K) = 0$ for all $a > 0$.*

We say that a subset of the complex plane is *locally of capacity zero* if all its bounded subsets have capacity zero.

THEOREM A.1.29 (H. CARTAN). *Let ϕ be subharmonic on a domain D of $\mathbf{C}$ and not identically $-\infty$. Then $\{z : z \in D, \phi(z) = -\infty\}$ is a G_δ-set which is locally of capacity zero.*

§2. Functions of Several Complex Variables

The list of results given below is the strict minimum we need. So it is a bit skeletal. If the reader needs more substance he must read [5,9,10] or any of the following books: L. Bers, *Introduction to Several Complex Variables*, New York, 1964; M. Fields, *Several Complex Variables and Complex Manifolds*, I and II, Cambridge, 1982; B. Fuks, *Special Chapters in the Theory of Analytic Functions of Several Complex Variables*, Providence, 1965; H. Grauert and K. Fritzsche, *Several Complex Variables*, New York, 1976; R. Gunning and H. Rossi, *Analytic Functions of Several Complex Variables*, Englewood Cliffs, 1965; S.G. Krantz, *Function Theory of Several Complex Variables*, New York, 1982; P. Lelong, *Fonctions analytiques et fonctions entières (n variables)*, Montréal, 1968; and R. Narasimhan cited in §1.

THEOREM A.2.1 (F. HARTOGS). *Let D be an open subset of $\mathbb{C}^n$ and suppose that f is separately holomorphic on D. Then f is holomorphic on D.*

We know that there exist open subsets D of $\mathbb{C}^n$ ($n \geq 2$) such that each holomorphic function on D can be extended holomorphically on a greater open set. So we are now interested in open sets which do not have this property.

We say that an open subset D of $\mathbb{C}^n$ ($n \geq 2$) is an *open set of holomorphy* if D is non-empty and if there exists a function f holomorphic on D which cannot be extended holomorphically on a greater open set: that is, for each $a \in \partial D$ there exists a neighbourhood V such that for any connected neighbourhood U of a, included in V, there is no holomorphic function F which coincides with f on a non-empty open subset of $U \cap D$.

If K is a compact subset of D we define the $H(D)$-*hull* of K by the following: $\hat{K}_D = \{x : x \in D, |f(x)| \leq \max_{u \in K} |f(u)|, \text{for every } f \in H(D)\}$. It is obvious that $K \subset \hat{K}_D$. Considering $f(z) = \exp(z|\xi)$, where $(z|\xi)$ is the scalar product of z with a vector $\xi \in \mathbb{C}^n$, we conclude that $\hat{K}_D$ is included in the convex hull of K. It is also obvious that $\hat{K}_D$ is closed in D, but in general it is not compact in D.

THEOREM A.2.2 (H.CARTAN-P.THULLEN). *Let D be a non-empty open subset of $\mathbb{C}^n$. Then the following properties are equivalent:*

(i) *D is an open set of holomorphy,*

(ii) *if K is an arbitrary compact subset of D then $\hat{K}_D$ is also a compact subset of D.*

COROLLARY A.2.3. *If D is a non-empty convex open subset of $\mathbb{C}^n$ then D is an open set of holomorphy.*

In particular open balls are open sets of holomorphy.

THEOREM A.2.4. *Let $(D_i)_{i \in I}$ be a family of open sets of holomorphy in $\mathbb{C}^n$. Then the interior of $\cap_{i \in I} D_i$ is also an open set of holomorphy in $\mathbb{C}^n$.*

THEOREM A.2.5. *Let D and D' be open sets of holomorphy respectively in $\mathbb{C}^n$ and $\mathbb{C}^m$ and let u be a holomorphic map from D into $\mathbb{C}^m$, that is all its components $u_1, \ldots, u_m$ are holomorphic on D. Then*

$$D_u = \{z : z \in D, u(z) \in D'\}$$

is an open set of holomorphy in $\mathbb{C}^n$.

COROLLARY A.2.6. *Let D be an open set of holomorphy in $\mathbf{C}^n$ and let $f_1, \ldots, f_k \in H(D)$. Then $D_f = \{z : z \in D, |f_i(z)| < 1, i = 1, \ldots, k\}$ is an open set of holomorphy in $\mathbf{C}^n$.*

It is well-known that an open subset U of $\mathbf{R}^n$ is convex if and only if $-\log \text{dist}(z, \partial U)$ is convex for $z \in U$. This suggests the following definition.

We say that an open subset D of $\mathbf{C}^n$ is *pseudoconvex* if it is non-empty and if $-\log \text{dist}(z, \partial D)$ is plurisubharmonic on D.

THEOREM A.2.7. *Let (D_k) be an increasing sequence of pseudoconvex open sets of $\mathbf{C}^n$. Then their union is pseudoconvex in $\mathbf{C}^n$.*

THEOREM A.2.8. *Let $(D_i)_{i \in I}$ be a family of pseudoconvex open subsets of $\mathbf{C}^n$. Then the interior of $\cap_{i \in I} D_i$ is pseudoconvex in $\mathbf{C}^n$.*

If K is a compact subset of the open set $D \subset \mathbf{C}^n$, we define the $P(D)$-*hull of* K, denoted $\tilde{K}_D$, by

$$\tilde{K}_D = \{z : z \in D, \phi(z) \le \max_{u \in K} \phi(u), \text{ for all } \phi \text{ plurisubharmonic on } D\}.$$

It is obvious that $K \subset \tilde{K}_D \subset \hat{K}_D$.

THEOREM A.2.9. *Let D be a non-empty open subset of $\mathbf{C}^n$. Then the following properties are equivalent:*

(i) D is pseudoconvex,

(ii) there exists ϕ continuous and plurisubharmonic on D such that the sets $D_c = \{z : z \in D, \phi(z) < c\}$ are relatively compact in D for all real numbers c,

(iii) the same property without requiring the continuity of ϕ,

(iv) if K is compact in D then $\tilde{K}_D$ is also a compact subset of D.

THEOREM A.2.10. *Let D be an open set of holomorphy in $\mathbf{C}^n$. Then D is pseudoconvex in $\mathbf{C}^n$.*

THEOREM A.2.11. *An open subset D of $\mathbf{C}^n$ is pseudoconvex if and only if there exists a plurisubharmonic function ϕ on D that goes to $+\infty$ when z goes to the boundary of D.*

THEOREM A.2.12. *Let $D_1 \subset \mathbf{C}^{n_1}, \dots, D_k \subset \mathbf{C}^{n_k}$ be pseudoconvex open sets and let $n = n_1 + \cdots n_k$. Then $D_1 \times \cdots \times D_k$ is pseudoconvex in $\mathbf{C}^n$.*

THEOREM A.2.13. *Let D be a pseudoconvex open subset of $\mathbf{C}^n$ and u be a bi-holomorphic mapping from D onto $D' \subset \mathbf{C}^n$. Then D' is pseudoconvex.*

The next result is a generalization of Theorem A.2.11 which is used in Chapter VII.

Let D be a non-empty open subset of $\mathbf{C}^n$. A real function ϕ defined on D is called a *vertical function for D* if, for any $a \in \mathbf{C}$ such that $\{(a, z'): z' \in \mathbf{C}^{n-1}\}$ intersects D, we have

$$\lim_{z' \to z'_0} \phi(a, z') = +\infty$$

whenever (a, z'_0) is a boundary point of D.

THEOREM A.2.14. *Let D be a non-empty open subset of $\mathbf{C}^n$. Then D is pseudoconvex if and only if there exists a vertical function ϕ for D which is plurisubharmonic on D.*

It is a classical result in the theory of convex sets (perhaps not well-known) that an open subset D of $\mathbf{R}^n$ is convex if and only if, for all a in the boundary of D, there exists an open neighbourhood V of a such that $V \cap D$ is convex. This is a very striking property because, at first sight, the definition of convexity is not a local one. For pseudoconvex open sets there is a similar result.

THEOREM A.2.15. *Let D be a non-empty open subset of $\mathbf{C}^n$. Then D is pseudoconvex if and only if, for every boundary point of D, there exists an open neighbourhood V of that point such that $V \cap D$ is pseudoconvex.*

The next result, along with Theorem A.2.17, is much used in Chapter VII.

THEOREM A.2.16. *Let D be a pseudoconvex open subset of $\mathbf{C}^n$. Then there exists $\phi \in \mathbf{C}^\infty(D)$ such that*

(i) *ϕ is strictly plurisubharmonic on D,*

(ii) *for all real numbers c, the sets $D_c = \{z: z \in D, \phi(z) < c\}$ are relatively compact in D and $D = \cup_{k=1}^\infty D_k$.*

By Theorem A.2.10 we know that open sets of holomorphy are pseudoconvex in $\mathbf{C}^n$. The converse problem, that all pseudoconvex open subsets of $\mathbf{C}^n$ are open sets of holomorphy, was settled by E. Levi in 1911. It was solved, for $n = 2$, by K. Oka in 1942, and for $n \geq 2$ by K. Oka, F. Norguet and H.J. Bremermann in 1953–1954.

By Theorem A.2.16 and the Behnke-Stein theorem, which asserts that the union of an increasing sequence of open sets of holomorphy is an open set of holomorphy, the problem is to prove that $D_c = \{z : z \in D, \phi(z) < c\}$ is an open set of holomorphy for $\phi \in C^\infty(D)$ which is strictly plurisubharmonic on D. In fact it is necessary to solve Cousin's problem. For example K. Oka proved the following lemma. Let D be a bounded domain in $\mathbf{C}^n$ and $\alpha, \beta \in \mathbf{R}$ such that $\alpha < \beta$, and suppose that $D_1 = \{z : z \in D, \operatorname{Re} z_1 > \alpha\}$ and $D_2 = \{z : z \in D, \operatorname{Re} z_1 < \beta\}$ are domains of holomorphy. Then D is a domain of holomorphy.

For $n = 2$, the case of interest to us, the proof of Levi's problem is not too difficult (see for instance the book by B. Fuks, pp. 194–215). For $n \geq 2$, many proofs have been given since 1955, using Oka's original method or functional analysis or partial differential equations (Ehrenpreis, Grauert, Narasimhan, Andreotti-Grauert, J.J. Kohn etc.).

Let D be an open subset of $\mathbf{C}^n$ and let $a \in \partial D$. We say that D has a *variety of support* at a if there exist a neighbourhood U of a and f holomorphic on U such that $\{z : z \in U, f(z) = 0\} \cap \overline{D} = \{a\}$.

It is well-known that convex subsets of $\mathbf{C}^n$ have hyperplanes of support at each boundary point. So this suggests the following general problems which are unsolved until today.

(a) Given a pseudoconvex subset D of $\mathbf{C}^n$, at which boundary points does D have varieties of support?

(b) For which classes of rather smooth pseudoconvex subsets of $\mathbf{C}^n$ is it possible to have varieties of support through all boundary points?

We give only a partial solution to (b) which is due to K. Oka. The original proof is rather complicated, particularly when $\operatorname{grad} \phi(\lambda_0, z_0) = 0$. There is a more elementary solution due to A. Zraïbi. Using Narasimhan's lemma (see the book by S.G. Krantz, p. 111) or the Levi-Krzoska theorem (see [0], pp. 157–162) it is even possible to extend this for $D \subset \mathbf{C}^n$ but, in fact, we do not need this general result.

THEOREM A.2.17. *Let D be an open subset of $\mathbf{C}^2$ and let ϕ be a C^2-strictly plurisubharmonic function on D such that $D_0 = \{(\lambda, z) : \lambda, z \in \mathbf{C}, \phi(\lambda, z) < 0\}$ is relatively compact in D. Then for all $(\lambda_0, z_0) \in \partial D_0$, either there exist $r > 0$ and h holomorphic on $B(\lambda_0, r)$ such that $h(\lambda_0) = z_0$ and $(\lambda, h(\lambda)) \notin D_0$ for $|\lambda - \lambda_0| < r$, or there exist $s > 0$ and k holomorphic on $B(z_0, s)$ such that $k(z_0) = \lambda_0$ and $(k(z), z) \notin D$ for $|z - z_0| < s$.*

REFERENCES

1. Aupetit, Bernard: *Propriétés spectrales des algèbres de Banach.* Lecture Notes in Mathematics **735**. Springer-Verlag, Berlin-Heidelberg-New York, 1979.

2. Bonsall, Frank F. and Duncan, John: *Complete Normed Algebras.* Ergebnisse der Mathematik und ihrer Grenzgebiete **80**. Springer-Verlag, New York-Heidelberg-Berlin, 1973.

3. Halmos, Paul R.: *A Hilbert Space Problem Book.* D. Van Nostrand, Princeton, 1967.

4. Hayman, W.K. and Kennedy, P.B.: *Subharmonic Functions. Volume I.* London Mathematical Society Monographs **9**. Academic Press, London-New York-San Fransisco, 1976.

5. Hörmander, Lars: *An Introduction to Complex Analysis in Several Variables.* North-Holland Mathematical Library **7**. North-Holland Publishing Co., Amsterdam-London, 1973.

6. Rickart, Charles E.: *General Theory of Banach Algebras.* The University Series in Higher Mathematics. D. Van Nostrand, Princeton, 1960.

7. Rudin, Walter: *Real and Complex Analysis.* Second Edition. McGraw-Hill Series in Higher Mathematics. McGraw-Hill Book Co., New York 1974.

8. Rudin, Walter: *Functional Analysis.* McGraw-Hill Series in Higher Mathematics. McGraw-Hill Book Co., New York, 1973.

9. Vladimirov, V.S.: *Methods of the Theory of Functions of Many Complex Variables.* M.I.T. Press, Cambridge, Mass., 1966.

10. Wermer, John: *Banach Algebras and Several Complex Variables.* Second Edition. Graduate Texts in Mathematics **35**. Springer-Verlag, New York-Heidelberg-Berlin, 1976.

AUTHOR AND SUBJECT INDEX

Universitext *(continued)*